Visual Analysis of Multilayer Networks

Synthesis Lectures on Visualization

Editors
Niklas Elmqvist, *University of Maryland*
David S. Ebert, *University of Oklahoma*

Synthesis Lectures on Visualization publishes 50- to 100-page publications on topics pertaining to scientific visualization, information visualization, and visual analytics. Potential topics include, but are not limited to: scientific, information, and medical visualization; visual analytics, applications of visualization and analysis; mathematical foundations of visualization and analytics; interaction, cognition, and perception related to visualization and analytics; data integration, analysis, and visualization; new applications of visualization and analysis; knowledge discovery management and representation; systems, and evaluation; distributed and collaborative visualization and analysis.

Design of Visualizations for Human-Information Interaction: A Pattern-Based
Framework
Kamran Sedig and Paul Parsons
2016

Image-Based Visualization: Interactive Multidimensional Data Exploration
Christophe Hurter
2015

Interaction for Visualization
Christian Tominski
2015

Data Representations, Transformations, and Statistics for Visual Reasoning
Ross Maciejewski
2011

A Guide to Visual Multi-Level Interface Design From Synthesis of Empirical Study
Evidence
Heidi Lam and Tamara Munzner
2010

Visual Analysis of Multilayer Networks

Fintan McGee, Benjamin Renoust, Daniel Archambault, Mohammad Ghoniem, Andreas Kerren, Bruno Pinaud, Margit Pohl, Benoît Otjacques, Guy Melançon, and Tatiana von Landesberger

ISBN: 978-3-031-01480-2   paperback
ISBN: 978-3-031-02608-9   ebook
ISBN: 978-3-031-00352-3   hardcover

DOI 10.1007/978-3-031-02608-9

A Publication in the Springer series
SYNTHESIS LECTURES ON VISUALIZATION

Lecture #12
Series Editors: Niklas Elmqvist, *University of Maryland*
                David S. Ebert, *University of Oklahoma*
Series ISSN
Print 2159-516X   Electronic 2159-5178

Visual Analysis of Multilayer Networks

Fintan McGee
Luxembourg Institute of Science and Technology (LIST), Luxembourg

Benjamin Renoust
University of Osaka, Japan

Daniel Archambault
Swansea University, United Kindom

Mohammad Ghoniem
Luxembourg Institute of Science and Technology (LIST), Luxembourg

Andreas Kerren
Linköping University, Sweden

Bruno Pinaud
University of Bordeaux, France

Margit Pohl
Vienna University of Technology, Austria

Benoît Otjacques
Luxembourg Institute of Science and Technology (LIST), Luxembourg

Guy Melançon
University of Bordeaux, France

Tatiana von Landesberger
University of Rostok, Germany
University of Cologne, Germany

SYNTHESIS LECTURES ON VISUALIZATION #12

ABSTRACT

The emergence of multilayer networks as a concept from the field of complex systems provides many new opportunities for the visualization of network complexity, and has also raised many new exciting challenges. The multilayer network model recognizes that the complexity of relationships between entities in real-world systems is better embraced as several interdependent subsystems (or layers) rather than a simple graph approach. Despite only recently being formalized and defined, this model can be applied to problems in the domains of life sciences, sociology, digital humanities, and more. Within the domain of network visualization there already are many existing systems, which visualize data sets having many characteristics of multilayer networks, and many techniques, which are applicable to their visualization. In this Synthesis Lecture, we provide an overview and structured analysis of contemporary multilayer network visualization. This is not only for researchers in visualization, but also for those who aim to visualize multilayer networks in the domain of complex systems, as well as those solving problems within application domains. We have explored the visualization literature to survey visualization techniques suitable for multilayer network visualization, as well as tools, tasks, and analytic techniques from within application domains. We also identify the research opportunities and examine outstanding challenges for multilayer network visualization along with potential solutions and future research directions for addressing them.

KEYWORDS

multilayer networks, visualization, network visualization, visual analytics, interaction, task taxonomy, attribute visualization, multivariate network visualization, evaluation

Contents

Preface

Although not presented using the concept of multilayer networks, the 1985 paper "Relational contents in multiple network systems" by Burt and Schøtt [1985] clearly brings it into play. The ideas they laid out in their seminal paper were generalized, further examined and extended in a series of papers by Renoust et al. [2013, 2014, 2015]. Renoust transferred the notion of relation multiplexity to document collections, and later applied his approach to any collection of homophily relations in any context. At the same time, Kivelä et al. [2014] were developing a formal framework unifying all preceding efforts to deal with heterogeneous, multi-source networks.

Many of the ideas, concepts, definitions, and overall content of this book were given a decisive boost within the FR-LUX bilateral research project BLIZAAR[1] between its writing early 2015 to its official end in 2019. Incidentally, the writing of a survey on the visualization of multilayer networks [McGee et al., 2019a] was a deliverable of this project. Other projects[2] involving the Bordeaux team with sociologists and law researchers investigating human-trafficking [Lavaud-Legendre et al., 2017] definitely contributed to mature the concept of multilayer network and some of its intimate requirements when it comes to building systems supporting their visualization and navigation.

The authors would also like to thank Schloss Dagstuhl and the participants of seminar #19061 on the "Visualization of Multilayer Networks across Domains" [Kivelä et al., 2019]. The seminar was a unique event gathering both researchers of the visualization field, complex systems theory, and a significantly large group of domain experts all familiar with the concepts and in demand of the visualization of multilayer networks. Without the support of Schloss Dagstuhl and the commitment of its participants, this book would have never come to be.

In October 2019, all authors of this book participated to the "Multilayer Network Visualization" workshop organized on the occasion of the IEEE VIS'19 week conferences held in Vancouver. We would like to thank the organizers of IEEE VIS'19 for their support for the workshop. One of the results of the workshop was to gather the authors around the project of writing this book.

[1]BLIZAAR was funded by the French national research agency (ANR), grant ANR-15-CE23-0002-01, and the Luxembourg National Research Fund (FNR), grant INTER/ANR/14/9909176. The project involved the Luxembourg Institute of Science and Technology (LIST), the University of Bordeaux, the center for Digital Research in European Studies (CVCE https://www.cvce.eu/), and EISTI (https://eisti.fr/).
[2]We acknowledge the support from University of Bordeaux through the 2015–2016 TETRUM grant, and from the GIP Justice for their financial support of the 2017–2019 AVRES grant number 217.03.30.05.

Finally, we would like to thank Niklas Elmqvist and all at Morgan & Claypool for supporting us in the production of this book.

Fintan McGee, Benjamin Renoust, Daniel Archambault, Mohammad Ghoniem,
Andreas Kerren, Bruno Pinaud, Margit Pohl, Benoît Otjacques, Guy Melançon,
and Tatiana von Landesberger
April 2021

Figure Credits

Figure 3.1   From B. Renoust, G. Melançon, and T. Munzner. Detangler: Visual analytics for multiplex networks. *Computer Graphics Forum*, 34(3):321–330, 2015. John Wiley and Sons. Used with permission.

Figure 3.2   From Renoust et al., 2015. John Wiley and Sons. Used with permission.

Figure 3.3   From Renoust et al., 2015. John Wiley and Sons. Used with permission.

Figure 3.4   From V. Yoghourdjian, T. Dwyer, K. Klein, K. Marriott, and M. Wybrow. Graph thumbnails: Identifying and comparing multiple graphs at a glance. *IEEE Transactions on Visualization and Computer Graphics*, 24(12):3081–3095, 2018. IEEE. Used with permission.

Figure 3.5   From M. Freire, C. Plaisant, B. Shneiderman, and J. Golbeck. ManyNets: An interface for multiple network analysis and visualization. In *Proc. of the SIGCHI Conference on Human Factors in Computing Systems, CHI'10*, pages 213–222, 2010. Association for Computing Machinery, Inc. Reprinted by permission.

Figure 3.6   From M. Rosvall and C. T. Bergstrom. Mapping change in large networks. *PloS One*, 5(1), 2010. Used with permission under Creative Commons license Attribution 4.0 International (CC BY 4.0).

Figure 5.5   From S. Ghani, B. C. Kwon, S. Lee, J. S. Yi, and N. Elmqvist. Visual analytics for multimodal social network analysis: A design study with social scientists. *Transactions on Visualization and Computer Graphics*, 19(12):2032–2041, 2013. IEEE. Used with permission.

Figure 5.6   From H. Yang, K. Tang, X. Liu, L. Xiao, R. Xu, and S. Kumara. A user-centred approach to information visualisation in nanohealth. *International Journal of Bioinformatics Research and Applications*, 12(2):95–115, 2016. Inderscience Enterprises Ltd. Used with permission.

Figure 7.1   From S. Grottel, J. Heinrich, D. Weiskopf, and S. Gumhold. Visual analysis of trajectories in multi-dimensional state spaces. *Computer Graphics Forum*, 33(6):310–321, 2014. John Wiley and Sons. Used with permission.

Figure 7.2   From A. J. Pretorius and J. J. Van Wijk. Visual inspection of multivariate graphs. *Computer Graphics Forum*, 27(3):967–974, 2008. John Wiley and Sons. Used with permission.

Figure 7.3   From M. Burch, M. Höferlin, and D. Weiskopf. Layered TimeRadarTrees. In *Proc. of the 15th International Conference on Information Visualisation, IV*, pages 18–25, 2011. IEEE. Used with permission.

Figure 7.4   From M. De Domenico, M. A. Porter, and A. Arenas. MuxViz: A tool for multilayer analysis and visualization of networks. *Journal of Complex Networks*, 3(2):159–176, 2015. Oxford University Press. Used with permission.

Figure 7.5   From I. Jusufi, A. Kerren, and B. Zimmer. Multivariate network exploration with JauntyNets. In *17th International Conference on Information Visualisation*, pages 19–27, 2013. IEEE. Used with permission.

Figure 8.1   From F. Schreiber, A. Kerren, K. Börner, H. Hagen, and D. Zeckzer. Heterogeneous networks on multiple levels. In A. Kerren, H. C. Purchase, and M. O. Ward, Eds., *Multivariate Network Visualization: Dagstuhl Seminar #13201*, Dagstuhl Castle, Germany, May 12–17, 2013, Revised Discussions, pages 175–206, 2014. Springer International Publishing. Used with permission.

CHAPTER 1

Introduction and Overview

Contents

1.1 INTRODUCTION

The emergence of multilayer networks as a concept from the field of complex systems provides many new opportunities for the visualization of network complexity, and has also raised many new exciting challenges. The multilayer network model recognizes that the complexity of relationships between entities in real-world systems is better embraced as several interdependent subsystems (or layers) rather than a simple graph approach. Despite only recently being formalized and well defined, this model can be applied to problems in the domains of life sciences, sociology, digital humanities, and more. Within the domain of graph visualization there already are many existing systems, which visualize data sets having many characteristics of multilayer networks, and many techniques, which are applicable to their visualization. Previously, simple graphs were often used to model relationships between entities in real-world systems. Such a graph, in its most basic form, is defined as a tuple $G = (V, E)$, where V denotes the set of vertices (or nodes) which represent entities, and E is a set of edges (or links), which are vertex pairs that represent the relationships between entities.[1] This approach may, however, be an over-

---

[1] Since this terminology heavily depends on the background of the authors writing the scientific articles we are building our book upon, we use interchangeably the terms *graph* and *network*, *nodes* and *vertices*, and *links* and *edges*, without any preference.

simplification of a much more complex reality that may be more accurately modeled as several interdependent subsystems (or layers), which motivated the development of the *complex networks* field (see Gao et al. [2012] and Kenett et al. [2015]). The concept of a multilayer network, as formulated in the seminal work of Kivelä et al. [2014], builds on and encompasses many existing network definitions across many fields, some of which are much older, e.g., from the domain of sociology, such as the work of Burt and Schøtt [1985], Moreno [1934], and Verbrugge [1979].

1.2   AN INTRODUCTION TO MULTILAYER NETWORK CONCEPTS

Layers are the fundamental concept for multilayer networks, and we will discuss them more formally along with other important concepts in Chapter 3. However, to introduce the concept we will use an initial illustrative example of a person's social networks.

People frequently use more than one social network platform, e.g., Facebook for their personal social network or LinkedIn for their professional one. Offline, "real life," social networks could also be considered, again with relations being either personal or professional. These networks can be considered independent, however, they can also be considered as layers in a multilayer network. The networks overlap as some people may be present across layers. Layers are in this case characterized by relationship type (either online/offline and personal/professional). A significant change in one network may implicitly correlate with or cause changes in another. For example, a change of employer will cause changes in both offline and online professional networks but in a different manner for each, and may cause slower, more gradual, changes in the personal offline/online social networks. To answer some questions, it may be necessary to also include employers or companies as entities of the network. This makes it possible to model explicitly person-company relationships, as well as person-person and company-company relationships. In this case, layers may be characterized by entity type (either person or company). Other definitions of layers are also possible, such as defining layers based on the relationship type, or based on the time period of interactions. When an edge is entirely within a layer (i.e., its start and end nodes are within the same layer), it is referred to as an **intra**-layer edge. When modeling layers some relationships may connect between entities in different layers. These edges between the layers are referred to as **inter**-layer edges. If the edge is connecting the same entity in different layers it is referred to as a *coupling* edge.

It is important to emphasize that layers do not reduce to some operational apparatus. The concept goes far beyond a simple intent to capture data heterogeneity. While it is true this notion is most of the time embodied as nodes and edges of a network being of different "types," its roots lie deeply in sociology research, such as the work of Burt and Schøtt [1985], Geard and Bullock [2007], and Lazega and Pattison [1999]. This notion is used to form questions and hypotheses, where layers can be considered as innermost, intermediate, or outer layers, as done by Lin [2008]. For instance, Dunbar et al. [2015] and Manivannan et al. [2018] consider networks similar to our introductory example, and examine to what extent online and offline layers in personal networks overlap.

While innermost and outermost layers are well established notions in sociology, the modeler is free to be "creative" when deciding what constitutes a layer (*dixit* Kivelä et al. [2014]). That is, the notion of a layer in a network emerges from and belongs to the domain under investigation. Consequently, when discussing the notion of layer, it is important to distinguish the real-world network from the mathematical network used to describe it. The mathematical network—a graph—is but an artifact through which we may hope to observe and ultimately characterize a phenomenon occurring on the sociological network. The definition of a layer is thus a characteristic of the multilayer system as a whole, defined either by a physical reality or the system being modeled. The notion of a layer naturally occurs when describing tasks performed by analysts; it can be mobilized to form exploration or browsing strategies (see Chapter 4 for a discussion on tasks).

1.2.1   A MORE FORMAL DEFINITION OF MULTILAYER NETWORKS

In order to define a multilayer graph in a more formal manner, we begin with the definition of a standard graph. A standard graph is often described by a tuple $G = (V, E)$ where V defines a set of vertices and E defines a set of edges (vertex pairs), such that $E \subseteq V \times V$.

An intuitive definition of a multilayer network first consists of specifying which layers the network nodes belong to. Given a set of layers L, with an individual layer being defined as $l \in L$, and given that we we allow a node $v \in V$ to be part of some layers and not others, we may consider nodes in a multilayer graph as pairs $V_M \subseteq V \times L$. Edges $E_M \subseteq V_M \times V_M$ then connect pairs $(v, l), (v', l')$. An edge is said to be *intra* or *inter*-layer depending on whether $l = l'$ or $l \neq l'$.

In our preceeding example, we would have $L = \{l, l', l'', \ldots\}$ where $l =$ Facebook friends, $l' =$ LinkedIn connections, $l'' =$ "real life" family-friends-acquaintances, etc.

There are many different types of network which fall under the umbrella of multilayer networks. If the set of nodes is disjoint across layers, i.e., each layer has a different set of nodes, nodes do not repeat, then the network has the same properties as what is known as a *node-colored* network. If networks have only inter-layer edges and no intra-layer edges, then we can consider them having the same properties as *bi-partite*, for two layers, or *multipartite/k-partite* networks with more than two layers. We will discuss in more detail the relationship between multilayer networks and other network types, such as dynamic and multivariate networks, in Chapter 2. We will discuss the concept of layers and their definition in more detail in Chapter 3.

1.2.2   MORE ADVANCED MULTILAYER NETWORK CONCEPTS

To fully understand multilayer networks, it is necessary to have more than just an understanding of the concept of a layer. To introduce the more advanced concepts we consider an example from the domain of life sciences, an application domain for multilayer network visualization that is further discussed in Chapter 2. Biologists frequently consider different sets of data when analyzing a biological entity. They may consider genes (genomics data), metabolites (metabolomic data), and proteins (proteomic data). Networks can be created modeling the relationships (links) between the entities (nodes) in each of these data sets. However, there can also be interactions between different entities in each of these data sets, so they should not be considered in isolation. Therefore, each of these networks can be considered as layers in a multilayer network, with the relationship between the data sets captured as inter-layer edges. It is the entity data type which characterizes each of the layers, or in multilayer network terminology we can say that the layers are part of the data type *aspect*.

However, data type may not be the only way to characterize the entities being modeled. The amount of data involved in modeling a biological organism is vast, and a biologist may only look at subsets of network data related to a specific biological function, often referred to as biological pathways. There may be multiple pathways of interest to the biologist, and these may overlap, and may also have relationships between entities across pathways. The notion of biological pathway can be used to characterize another layering of the data. This can be considered a second aspect in the multilayer network model. Figure 1.1 shows an illustrative example of this network and its two aspects. If we consider both aspects there are nine layers, each layer characterized by the intersection of aspects. It is not necessary to consider all aspects when examining a layer. If we only consider the data type aspect, each of the genomic, proteomic, and metabolomic layers will contain the nodes and relationships from all of the biological pathways.

Additional aspects could also be defined, for example, if the biological data contains time information, that may also be considered an aspect. While multiple aspects are a possibility for multilayer network data sets, it is not a requirement. A multilayer data set may be defined by a single aspect, which categorizes multiple layers.

The definition of aspects depends on the application domain and what is useful for the person using the multilayer networks to answer a question. See Table 1.1 for a sample list of aspects and layers extracted from some of the work cited within this book. Kivelä et al. [2014] provide further examples in their extensive list of multiplex data sets and their associated layers.

Incidentally, Wehmuth et al. [2016] propose an alternative definition they call Multi-Aspect graphs where they formally define what can be considered as an aspect. Unsurprisingly, they also form a network where nodes are defined using Cartesian products collecting multiple values into a single entity. The authors describe MultiAspect graphs as forming a generalization of the multilayer network of Kivelä et al. [2014]. Reconciling these different approaches is beyond the scope of this book.

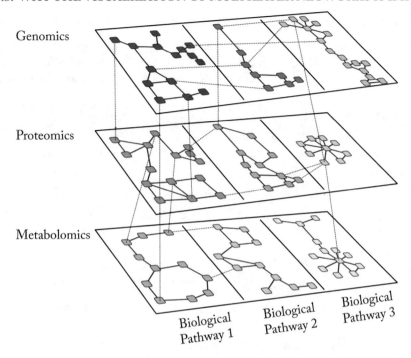

Figure 1.1: A purely illustrative example of multilayer data in the context of biology. A layer can be described by the type of data as a first aspect (genomic, proteomic, or metabolomic), and biological pathway being represented as a second aspect.

1.3   WHY THE VISUALIZATION OF MULTILAYER NETWORKS IS IMPORTANT

1.3.1   THE UBIQUITY OF MULTILAYER NETWORKS

Examples of multilayer networks can be found across a wide range of domains. In the domain of biology, where multilayer networks have been explicitly recognized as promising for biological analysis, see Gosak et al. [2017], the multilayer framework is a natural fit for the so-called "*omics*" layers, related to the analysis of biological entities such as proteins, metabolites, genes, etc. (see previous section). Multilayer networks are frequently used in epidemiology to analyze the spread of disease as done by Pastor-Satorras et al. [2015], Saumell-Mendiola et al. [2012], and Wang and Xiao [2012].

Within sociology (in a broad sense, including fields such as criminology, for instance) a multilayer approach can be used to analyze criminal networks, see Bright et al. [2015] and Lavaud-Legendre et al. [2017]; social networks in schools, see Crnovrsanin et al. [2014]; business organizations, see Lazega and Pattison [1999]; academic funding and collaboration

Table 1.1: Examples of aspects and layers, extracted from papers cited in later chapters

Aspect Description	Layer Definition	Domain
Social entity	People, societies/organizations	Information visualization Renoust et al. [2015]
Social relationship	Friendship, aggression	Social networks Crnovrsanin et al. [2014]
Word relationship Year of publication	Hyponym, homonym [1974, . . . , 2004]	Information visualization Hascoët and Dragicevic [2012]
Infrastructure connection	Air connection, train connection	Physics Halu et al. [2014]
Transport mode	Air, rail, ferry, coach	Scientific data (Transportation) Gallotti and Barthelemy [2015]
"Omics" Entity	Gene, protein, protein structure	Biology Pavlopoulos et al. [2008]
Historical correspondences	Letter, letter sender, letter receiver, cited book	Historical network research van Vugt [2017]
Building layout	Arrangement of house spaces	Robot Control Algorithms Ślusarczyk et al. [2017]

networks, see Ghani et al. [2013] and Lavaud-Legendre et al. [2017]; and general social network analysis, see Dickison et al. [2016a] and Geard and Bullock [2007]. Indeed, the notion of many relationships between individuals, often called *multiplex* relationships, is seminal in sociology and one could argue that it already was present in the sociograms introduced by Moreno [1934]. The notion is central in the work of Burt and Schøtt [1985] where the challenge is to somehow simplify multiplex relationships, and consolidate and substitute them for relationships involving a smaller number of relation types to ease the analysis of the network.

Within the field of digital humanities a multilayer approach can be seen in the analysis of related documents and extracted entities, see McGee et al. [2016]; exploration of archaeological data, see Dunne et al. [2012]; and analysis of events in the media, see Sluban et al. [2016]. Examples can also be found in transportation networks and other civil infrastructure research in the work of Cardillo et al. [2013], Derrible [2017], and Ducruet [2017].

In the field of network visualization many systems visualize data sets having many characteristics of multilayer networks, albeit under a different title. Multiplex, e.g., Cardillo et al. [2013], Renoust et al. [2015]; heterogeneous, Dunne et al. [2012], Schreiber et al. [2014]; multimodal, Ghani et al. [2013], Heath and Sioson [2009]; multiple edge set networks, Crnovrsanin et al. [2014]; multi-edge, Parikh et al. [2012]; multi-relational, Cai et al. [2005]; interdependent networks, Gao et al. [2012]; interconnected networks, Saumell-Mendiola et al. [2012], and networks of networks, Kenett et al. [2015]. These are among the many names given to various types of data that are encapsulated by the multilayer network framework of Kivelä et al. [2014], which has emerged from the field of complex networks, a sub-domain of the field of complex systems.

More details and examples of application domains and related network visualization concepts are provided in Chapter 2.

1.3.2   HANDLING ADDITIONAL COMPLEXITY

Multilayer network data is inherently more complex. Indeed, as already described, the complexity of the underlying data is a motivation for choosing a multilayer approach. If this complexity is ignored and a more simple graph model is used, an analyst may miss out on important insights about the system being modeled. Many of the various graph models that fall under the framework of multilayer networks have been adapted to handle an additional level of complexity not handled by simple graph models (e.g., multiple node types for heterogeneous or multimodal networks, or multiple edge types for multiplex or multi-edge networks). A multilayer approach can also give different perspectives on large complex data sets through various approaches to defining layers. For example, a large social network data set may be better understood as a set of ego networks, where each layer represents the set of entities related to individuals of interest. A multilayer approach can also be used to relate seemingly independent data sets in a unified manner and provide new insights, e.g., Bianconi [2014] give the example of a multilayer network consisting of power-grid infrastructure and internet infrastructure where failures can cascade from one layer to another.

Multilayer networks thus offer a comprehensive framework allowing complex network data to be better understood and analyzed. However, more importantly, the intrinsic utility of the framework resides in the notion of layer used at its core, as an artifact bridging the gap between data and domain concepts. The notion of a layer is broadly used in various areas of science and culture—from image editing or CAD to geology, archaeology, graphical arts, and even molecular gastronomy—to express the common idea that an object of interest is composite and obtained by adding layers conveying the different facets of the object. An example is the notion of social stratification, see Parsons [1940], that indirectly refers to the notion of layers, with social actors being part of different strata according to various criteria. The work of Manivannan et al. [2018], focusing on a social network inferred from multichannel communication, in a sense demonstrates that the social network cannot be reduced to any one of its layers and is more than their sum. De Domenico [2018] gives another example of how domain layers naturally map onto data layers, similarly to the multilevel perspective on social networks developed by Lazega and Snijders [2015].

The literature proposes theoretical models guiding the creation and analysis of visualization systems with a clear intention to gain control over the gap between the different conceptual levels, e.g., Brehmer and Munzner [2013], Meyer et al. [2012], and Munzner [2009]. In terms of designing visualization approaches for multilayer networks, the holistic character of layers becomes an asset.

1.3.3   NEW CHALLENGES AND OPPORTUNITIES IN VISUALIZATION

The additional complexity of multilayer networks gives rise to new challenges and opportunities for research in visualization. The notion of a layer provides a new entity to be visualized and considered for analysis. The visualization of its constituent nodes and entities is also more complex as relationships with other layers may also need to be considered. The inclusion of multivariate data (i.e., attributes) is also more challenging in a multilayer context. The complexity of the networks and the diversity of user tasks during analysis, also lead to challenges when evaluating the effectiveness of new techniques. We describe these challenges and more in the upcoming chapters.

1.4   BOOK MOTIVATION AND STRUCTURE

1.4.1   WHO IS THIS BOOK FOR?

In this synthesis lecture, we provide an overview and structured analysis of contemporary multilayer network visualization. This is not only for researchers in visualization, but also for those who aim to visualize multilayer networks in the domain of complex systems, as well as those solving problems within application domains.

1.4.2   THE GOALS OF THIS BOOK

Initial steps have been made toward consolidating the work on visualization of multilayer networks from domains outside of the information visualization field, see the work of De Domenico et al. [2015] from the domain of complex systems, or from the domain of social networks the work of Dickison et al. [2016a], based on the complex systems paper of Rossi and Magnani [2015]. The goals of this book are to provide the full range of our target audience with the fundamental knowledge they need to be successful in the visual analysis of multilayer networks. We would like also to guide and motivate those who wish to push back the boundaries of the field. To this end, we discuss the ubiquity of these networks, the nature of layers as an entity, and the related end user tasks. We describe techniques for visualizing and interacting with layers, their constituent nodes and edges, and related attributes. We also examine the evaluation of multilayer network visualizations, to validate that chosen approaches and techniques are effective. Throughout this book, we not only discuss the current state of the art of the visual analysis of multilayer networks allowing readers to understand the available techniques and approaches, but also the outstanding research challenges to motivate new innovations. We hope our conclusions will guide visualization and complex systems researchers, as well as domain practitioners, in making new advances to the visual analysis of multilayer networks.

1.4.3   BOOK STRUCTURE

There are many different perspectives to multilayer network visualization as a field. We begin with a discussion on multilayer networks across application domains and the related concepts from the field of network visualization. Each of the subsequent chapters focuses in detail on one specific topic, discussing the techniques and approaches that are currently used, as well as identifying open research challenges and opportunities for researchers to push forward the state of the art. Below we give an overview of each chapter.

Multilayer Networks Across Domains

Multilayer networks are a topic that is rapidly increasing in popularity. In Chapter 2, we examine the wide range of domains which provide data and research challenges suitable for multilayer visualization. We also examine the many related concepts in terms of types of network structure as well as related topics in the visualization literature.

The Layer as an Entity

The concept of layers is unique to multilayer networks: layers bring a higher level of abstraction, and interesting structures can be inferred from their organization. In Chapter 3, we investigate how layers naturally appear in different domain considerations. We further study how layers can be defined and how aspects can help identify relevant layers of interest. Considering layers as entities offers interesting opportunities for visualization and manipulation, and many new challenges for interaction.

Tasks for Multilayer Networks

Following the detailed definition of a layer, layer related tasks are considered as fundamental as node and edge related tasks in simple graphs. In Chapter 4 we propose a task taxonomy for multilayer network centered on the use of layers. The taxonomy proposes task categories for cross layer tasks, layer structure manipulation and layer comparisons. We also present how existing visualization task taxonomies could be used or enhanced to handle multilayer networks.

Visualization of Nodes and Relationships Across Layers

While layers are a defining characteristic of multilayer networks, visualizing their constituent nodes and edges as well as the relationships across layers is key to the successful visual analysis of multilayer networks. In Chapter 5, we examine the wide range of techniques available for visualizing the nodes and relationships that make up a multilayer graph and its layers. We examine different approaches to encoding entities, and layout of entities with respect to their layers, and the visualization of edges.

Interacting with and Analyzing Multilayer Networks

The standard way of producing an information visualization is first to analyze the network and compute measures. Then, users can exploit the visualization, and even dig further with the support of carefully designed interactions. In Chapter 6, we first present user interaction approaches designed for or adaptable to multilayer networks. Then, we dive into analytics measures, where many of the existing methods for standard network have already been adapted to the multilayer case.

Attribute Visualization and Multilayer Networks

Multilayer networks may have different types of attributes associated with the network nodes and/or the edges. In Chapter 7, we motivate the importance of those attributes for the analysis of multilayer networks and provide an overview of the various attribute types. Such variations range from numeric, categorical, and temporal attributes, to specific attributes that are derived from so-called graph metrics on a per layer basis. Finally, we highlight the most important challenges and recent research opportunities/trends within the area of attribute visualization for multilayer networks.

Evaluation of Multilayer Network Visualization Systems and Techniques

Multilayer networks are very complex, therefore, their interpretation is a very challenging process. Human users may have serious difficulties to make sense of such visualizations because they will be overwhelmed by an abundance of information. Multilayer network visualizations should be designed appropriately to support the human user. Human users are able to detect complex patterns in visualizations easily and quickly. On the other hand, their attention and their short term memory is limited. In Chapter 8, we discuss tentative recommendations to take the specific character of human information processing into account when visualizing multilayer networks.

Conclusions

Within this book, we discuss existing techniques and a range of open challenges and research opportunities, related to many aspects of multilayer network visualization. In Chapter 9, we provide summary conclusions for each chapter.

CHAPTER 2

Multilayer Networks Across Domains

Contents

This chapter discusses the widespread popularity of multilayer networks (see Figure 2.1) and the potential impact that the development of effective visualizations can have on their wide range of applications. We also position the visualization of multilayer networks with respect to existing techniques in the field of network visualization—including k-partite and multivariate graphs—as well as highlight application domains that visualization experts may not be aware of. Throughout the literature on multilayer networks, from the early works of Moreno [1934] to more recent reviews such as that by Aleta and Moreno [2019], we emphasize the variety of application domains that multilayer network analysis techniques can be applied to. Specifically, we group applications into the domains of life sciences, social sciences, economics and finance, sustainability, digital humanities and multimedia, and finally, infrastructure and engineering. This list is not exhaustive, but clearly this range of domains is broad enough to motivate further research in the visualization of multilayer networks.

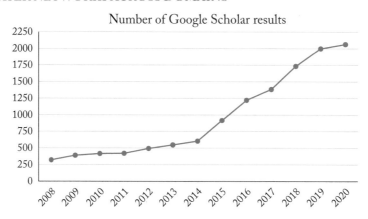

Figure 2.1: The number of results returned, per year, when searching for the term "Multilayer Networks" using Google Scholar. The uptick in 2014 coincides with the publication of Kivelä et al. [2014].

2.1   THE EMERGENCE OF MULTILAYER NETWORKS

The characterization of relationships between individuals is a core part of the discipline of Sociology. One could argue that the notion of multiple types of relationships was already studied in the early days of the discipline. Moreno was researching the sociometric structures of *sympathy*, *antipathy*, and *indifference* among children when he proposed his famous sociograms, see Moreno [1934], which may be the first historical attempt at visualizing multilayer networks. About half a century later, Burt and Schøtt [1985] address directly the challenge of studying the *relation content* in what they call *multiple networks*. Their aim was to somehow simplify multiplex relationships, consolidate and substitute them for relationships involving a smaller number of relation types to ease the analysis of the network.

More recently, Borgatti et al. [2009] defined a typology of the different ties studied in social network analysis, namely: *similarities* (or homophily) for individuals sharing characteristics, i.e., location, membership, attribute, etc.; *social relations* when they are explicit, such as family, friendship, coworkers, etc.; *interactions* of all sorts, e.g., shake hands, collaborated with, phoned to, etc.; and *flows*, capturing any kind of transfer, e.g., money, resources, information, etc.

Multilayer networks have recently attracted a growing attention from the complex systems domain, a physics and mathematics community, since the European project MULTI-PLEX [Caldarelli, 2012], and later when they were formally defined in the seminal paper by Kivelä et al. [2014].

The enduring nature of this interest can be seen by the continuing popularity of multilayer networks as a review topic within the physics community, e.g., Aleta and Moreno [2019]. Aside from the complex systems community, much early work using the fundamental concepts of

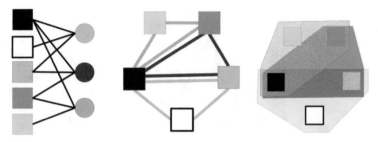

Figure 2.2: Equivalences exist between different models. Three models encode the same data set differently: a 2-partite model (left), a multilayer model (center), and a hypergraph model (right). All square nodes are the same entities in the three models. The relationships are encoded differently. The blue, green, and red round nodes in the bipartite network (left) each corresponds to an edge layers (or edge color) of the multilayer networks in the center, and to a hyperedge in the right.

multilayer networks was also seen in the fields of sociology, such as the work of Burt and Schøtt [1985], and economics, e.g., Snyder and Kick [1979].

The need for analysis of different levels of complexity is nothing new, and gave rise to different related network concepts that the visualization community has been tackling already. We now briefly introduce fundamental and recent works on related network visualization concepts, before we discuss the different application domains of multilayer networks.

2.2   RELATED NETWORK VISUALIZATION CONCEPTS

In this section, we review related graph models typically used for visualization and their differences and similarities to multilayer networks.

2.2.1   *K*-PARTITE GRAPHS

Recall that a bipartite graph is made of two disjoint sets of vertices so that no two vertices belonging to the same set are connected. Bipartite graphs can be considered as a case of multilayer networks with two layers and only inter-layer edges (see Figure 2.2, left). The two-mode (i.e., node type) nature of bipartite graphs results in analytics that are different to those of single-mode graphs [Borgatti and Everett, 1997].

Bipartite graph concepts are sometimes extended into k-partite (sometimes also referred to as n-partite, multipartite, or multimodal) graphs, although in practice many of the two-mode restrictions associated with bipartite graph are not fully retained. Systems which model bipartite cases and extensions of bipartite cases, such as the multimodal networks of Ghani et al. [2013], and the academic network analyzed by Shi et al. [2014], can be considered instances of multilayer

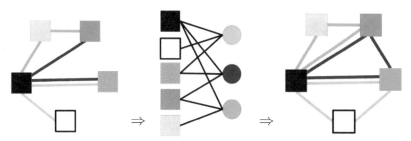

Figure 2.3: It might be tempting to model multilayer networks with multipartite networks only, however, there exist multilayer networks that do not have any equivalence. In this example, the original network (left) once transformed (centered) would imply the existence of edges (right) that were not present originally.

networks. In this case it is possible to also make use of bipartite analytics (e.g., adapted centrality metrics) to better understand their network structure.

Bipartite networks can be reduced to single mode networks via projection on one mode (i.e., all nodes of a specific type). Such an operation may be used to also define a layer in a multilayer network, if the projection results in a layer that reflects the reality of the system being modeled—although one should be careful not to introduce bias, as illustrated in Figure 2.3.

Bipartite and k-partite networks have been largely investigated to study the nature of social networks, for which Borgatti and Everett [1997] and before them Breiger [1974] are largely recognized. Often the different modes are considered separately [Lambiotte and Ausloos, 2006], while other works (e.g., Renoust et al. [2014]) exploit one mode to form multiple layers of a multilayer network (see Figure 2.2). Bipartite graphs may also correspond to incidence graphs (or Levi graphs from Levi [1942]) of hypergraphs. As a consequence, multilayer network modeling also extends to hypergraphs (see Figure 2.2), which are used to model numerous abstract problems such as in Gallo et al. [1993], but also social network analysis (e.g., McPherson [1982]). Two classical applications include database modeling (dependencies are modeled as hyperedges in Maier [1980]) and transportation systems (transportation lines correspond to hyperedges in Nguyen and Pallottino [1989]).

2.2.2   MULTIVARIATE GRAPHS

Multivariate graphs, as described by Kerren et al. [2014b], are those in which nodes or edges carry attributes or properties (illustrated in Figure 2.4). As described by Schreiber et al. [2014], there is a relationship between multivariate graphs and multilayer graphs. Some variables or attributes in a multivariate data set often serve the purpose of distinguishing nodes and edges that belong to different layers, e.g., the type of social network platform when analyzing a professional and personal social networks of an individual as a multilayer network. More recently, Nobre et al. [2019] provide a survey of multivariate network visualizations from a visualization design

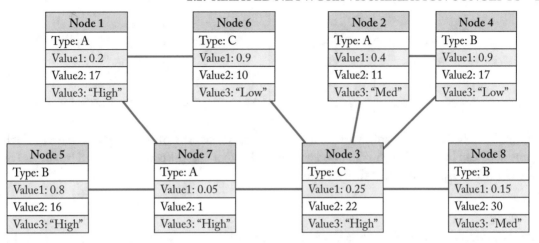

Figure 2.4: An illustrative multivariate graph, where each data node contains multiple attributes. Node attributes could be used to divide the network into layers. Defining layers by node type in this example would result in three layers, although that may not make sense for the system being modeled, as there would be no edge within the layers of nodes of type B and C.

perspective in which they distinguish between layouts, view operations, layout operations, and data operations. There are also multivariate visualization applications, such as that of Pretorius and Van Wijk [2008], that define their graph as having discrete sets, which can be considered analogous to defining layers. However, in the majority of cases, research into multivariate visualization lacks the *a priori* definition of a layer capturing a physical or conceptual reality related to the system being modeled.

Data sets can be considered *faceted* when multivariate data items are grouped in multiple orthogonal categories. Originally used as an approach to search and browse large data stores and text corpora (e.g., Cao et al. [2010] and Smith et al. [2006]), later work extended the faceted approach to include relationship visualization (e.g., Lee et al. [2009] and Zhao et al. [2013]). Data sets can have many different facets such as spatial and temporal frames of reference, or multiple values per data item and as such can be considered multifaceted. Visualizations for multifaceted data are those which show more than one of these facets simultaneously. Hadlak et al. [2015] provide a survey of multifaceted graph visualization techniques and discuss four common facets of network structure considered in network visualization, and their composition: partitions, attributes, time, and space. These facets may be considered to be very similar to instances of the aspects defined by Kivelä et al. [2014]. However, facets can be considered as different ways of exploring a single data set, which is unsurprising given the origins of a faceted visualization. The techniques described are still very useful for developing approaches for visualizing layers, particularly where the layer type matches the selected faceted categories of Hadlak et al. [2015]. However, faceted network visualization approaches do not meet all the

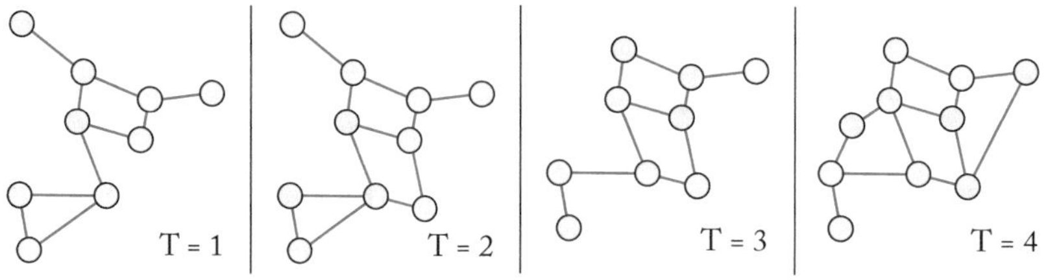

Figure 2.5: A dynamic graph, with four time slices, where structure changes over time. Each time slice can be intuitively understood as a layer. Further insight could be gained by the use of an additional aspect to define further layers.

needs for multilayer network visualization. While multilayer networks may use notions similar to these facets to characterize layers, multilayer network visualization also focuses on the inter-actions between layers and the role of layers in the network as a whole, while also supporting the concept of duplicated nodes and edges across layers.

2.2.3   DYNAMIC GRAPHS

Dynamic graphs are graphs whose structure (nodes and edges) and/or associated attributes may change over time. Analysts are often interested in comparing the state of the network at different points in time. Within the domain of complex networks, Boccaletti et al. [2014] consider the dynamics of multilayer networks. In many cases time slices of a dynamic (or temporal) network are simply mapped to layers. The notion of dynamic networks is also mentioned by Kivelä et al. [2014], who note that they can be considered as a type of multilayer network. A set of dynamic time slices can be considered layers in an aspect representing time. As multilayer networks can have multiple aspects, a temporal aspect might be just one of many. In their report on dynamic network visualization, Moody et al. [2005] explain the importance of "multiplicity" in social networks, i.e., the overlap of types of relations. In particular, they point out that linking relational timing to tie types allows better investigation of social dynamics. A recent survey of dynamic graph visualization techniques was provided by Beck et al. [2017], but does not consider layers in any context other than that of a hierarchical graph. The design space of dynamic network visualization is large and contains distinct approaches many of which relate to the change of structure between time-slices, see Kerracher et al. [2014]. This design space is a fertile ground for those seeking interesting visualization techniques to apply to the case of multilayer networks, particularly where the layers frequently contain the same nodes. One interesting example is the alluvial diagrams used by Rosvall and Bergstrom [2010].

2.3 MULTILAYER NETWORKS IN APPLICATION DOMAINS

Across all of the application domains mentioned in Section 1.3.1, advances in sensors, scientific equipment, online social networks, data collection and dissemination, and technology mean that researchers have access to more data than ever. This wealth of complex data is often best understood as a multilayer network model, and each of the different domains show particular characteristics that can be captured by multilayer networks, as discussed by Škrlj and Renoust [2020].

2.3.1 LIFE SCIENCES

Within biological network visualization there are many contexts in which a multilayer network approach may be beneficial, see Gosak et al. [2017]. Biologists have access to more genomic, proteomic, and metabolomic data than ever before, allowing for the construction of complex multilayer models of intricate biological processes. Interactions taking place within the genomic, proteomic, and metabolomic levels (and other so-called "*omics*" layers) can be modeled as individual networks, but interactions also occur between elements sitting in different omics levels within a larger biological system, where the aspect characterizing the layer is the node type, as seen in the work of Cottret et al. [2010]. This corresponds to the increasingly popular topic of systems/integrative biology, where the challenge consists in understanding the interplay and the cascade of effects taking place at the different levels of the biological system at hand, as discussed by Gehlenborg et al. [2010] and Kuo et al. [2013]. Within biotechnology and life sciences literature, it is recognized that an "*integrated omics analysis*" is preferable to analyzing the omics data sets independently, per de Carvalho et al. [2019]. Integrating multiple sources of omics data paints "*a more comprehensive picture,*" as pointed out by Aguiar-Pulido et al. [2016]. The multilayer network model lends itself to this distinction.

A prominent task for biologists analyzing biological pathways consists in comparing a species-specific pathway to a reference pathway [Murray et al., 2017], and in this specific case, species type can be considered a defining aspect for a layer. In medicine, deep phenotyping, see Delude [2015], where details about the manifestations of a disease are linked to omics layers is another domain of life sciences where the multilayer networks approach is very relevant. Fagherazzi [2020] describes the emerging next step in this field, which aims to achieve an even more comprehensive analysis by combining the phenotype and omics layers (as in deep phenotyping) with the digital footprint of the patients typically collected from wearables.

2.3.2 SOCIAL SCIENCES

Sociologists have shown an early interest in multilayer networks, in works such as that of Burt and Schøtt [1985], but also Podolny and Baron [1997], who brought one of the first multilayer-exclusive network measure (*multiplexity*). Gould [1991] applied multilayer networks to combine different social groups and localization in the investigation of the Paris Commune insurgency of 1871.

The multilayer network plays an important role in the forming and lasting of human social structure. This is highlighted in the seminal work of McPherson et al. [2001] who discuss the emergence of ties in social systems. Their work shows how similarity of people, i.e., *homophily*, can be strong driver to the formation of ties, that are also durable in a dynamic system. After investigating social ties in a multilayer manner, they argue for further research: *"in the impact of multiplex ties on the patterns of homophily; [and] the dynamic of network change over time through which networks and other social entities co-evolve."* Later research by Dow [2007] has also shown the relevance of multilayer networks for the transmission of homophily. Layer entanglement was explored by Renoust et al. [2014] as a means to capture the link between homophily and multiplexity, and successfully displays such a link when comparing social multilayer networks to other types of network in Škrlj and Renoust [2020].

Even when not explicitly categorized as multilayer, data sets within social network analysis frequently contain multiple types of edges. The purpose of the multiple edge sets may be to support examining the different types of relationships between people, as seen in the work of Crnovrsanin et al. [2014] (and also in much earlier work such as Lazega and Pattison [1999]), or to support multiple types of nodes, e.g., modeling a citation network containing researchers, institutions, and publications in Ghani et al. [2013]. Within social sciences, there are also contexts in which many networks may be compared to one another. One example is the examination of social networks produced as a result of cell phone activity, as done by Freire et al. [2010]. The contemporary use of multiple online social networks provides a vast amount of data, allowing for complex social multilayer networks to be built, that may help sociologists gain deeper insight.

Other fields such as food microbiology, have adopted social network analysis techniques, and applied them to understand problems such as the spread of diseases. For example, Crabb et al. [2017] use social network analysis techniques to understand the spread of salmonella in a large poultry farming enterprise. Different networks are generated based on contact between different types of entities. From a multilayer perspective, contact between entities can be considered an aspect, with the entity types defining the different layers. Toledo et al. [2020] use such a multilayer approach to capture the structure of criminal networks.

2.3.3   ECONOMICS AND FINANCE

The field of economics was an early adopter of multilayer network modeling. Snyder and Kick [1979] study an international network composed of ties very close to those defined by Borgatti et al. [2009], i.e., of trade flow, military intervention, conjoint relations, and treaty memberships. Nemeth and Smith [1985] also focus on the trade network from a multilayer perspective. Multilayer layer modeling goes beyond the structure of trading, such as modeling the whole infrastructure of a banking system in Bookstaber and Kenett [2016]. This is close to the spirit of Bargigli et al. [2015], who have highlighted the multiplex structures of banks. The work of Bazzi et al. [2016] combines temporal and multilayer networks to find communities of financial assets, while Biondo et al. [2017] investigate contagion in financial multilayer networks. Structural aspects of multilayer relationship between corporations and banks have also recently been pointed out by de Jeude et al. [2019] and Luu and Lux [2019]. Cryptocurrency can also be considered a topic of financial multilayer analytics, since the distributed nature of Blockchain is by its very essence forming networks (see Akcora et al. [2017]). Thanks to this characteristic, Bertazzi et al. [2018] model Bitcoin-OTC, a peer-to-peer market distributing Bitcoins, with multilayer networks to observe the prototypical behaviors linked to trust and reputation.

2.3.4   SUSTAINABILITY

Multilayer networks are also appropriate to study the sustainable management of natural resources. For instance, Geier et al. [2019] propose using them to better understand the interactions between three layers: governance, the people, and the resources.

From a broader perspective, the 17 Sustainable Development Goals (SDGs) adopted by the UN[1] and the 169 related targets identified in the 2030 Agenda for Sustainable Development can be modeled as a multilayer network in which every SDG can be seen as a layer where various stakeholders somehow intervene to contribute to several targets. Several studies (e.g., Pradhan et al. [2017]) have highlighted the complex relationships among some SDGs, i.e., across layers. Furthermore, the positive or negative impact on the SDGs of game changing technologies like Artificial Intelligence (see Vinuesa et al. [2020]) could also be integrated as a component of multilayer networks. The same statement applies to the relationship between the SDGs and major societal challenges like action on climate change (e.g., Nerini et al. [2019]). Although they have not been very much applied so far in this domain, multilayer networks seem thus to be a relevant approach to better understand interactions between all elements that are considered when reaching agreement over objectives for humanity.

2.3.5   DIGITAL HUMANITIES AND MULTIMEDIA

Digital humanities fields, such as digital cultural heritage, archaeology, and data journalism, have made already significant use of multilayer network modeling, for example, the work of Dunne et al. [2012], McGee et al. [2016], Müller et al. [2017], Sluban et al. [2016], and Van Vugt [2017]. This is supported by digital access to source texts and natural language processing techniques, such as Named-Entity Recognition and Topic Modeling, which allow for vast data sets to be built [McGee et al., 2016]. This enables the investigation of co-occurrence relationships between people, locations, organizations, as well as other entities forming multilayer networks. These multilayer networks may reveal insightful interaction patterns (see Renoust et al. [2014]). The sheer amount of multimedia programs across many channels and time slots also allows the creation of many co-occurrence networks of multiple aspects that may be investigated over time (Ren [2019], Ren et al. [2018a], and Renoust et al. [2016]). Beyond textual analysis, multi-

---

[1]https://sustainabledevelopment.un.org

layer networks have also been used to model stories across modalities, such as combination of script analysis, with face detection, and scene captioning (e.g., Mourchid et al. [2019]), or to enable a search engine in news (by Ren et al. [2018b]), or historical documents such as art pieces (from Garcia et al. [2020]) and statues [Renoust et al., 2019a].

2.3.6   INFRASTRUCTURE AND ENGINEERING

Modern vehicles often provide a wealth of telemetry and information that can be used to model transportation networks. These networks can also be combined in the framework of multilayer networks. For example, Halu et al. [2014] model the air and rail transportation networks of India as layers in a multilayer network. The work of Gallotti and Barthelemy [2015] is another good example. Parmentier et al. [2019] have investigated multilayer stream graphs—adding a temporal dimension to the multilayer network model—to explore the infrastructure network of the world's air traffic across different companies. In addition, a multilayer approach makes it possible to investigate higher order complex systems with strong dependency between layers, be it a transportation network or a computer network (see Kurant and Thiran [2006]).

The Internet and its associated infrastructure provide vast amounts of data about themselves and can be modeled as multilayer networks, as done early on by Cetinkaya and Knightly [2004] and Kurant and Thiran [2005]. More recently, Reis et al. [2014] describe the power grid and the Internet as separate interdependent layers in a multilayer infrastructure network. Derrible [2017] focuses on urban infrastructure systems, highlighting the necessity to adopt an integrated approach to urban planning by taking into account the interplay between multiple networks, such as transportation networks, energy networks, telecommunication networks, water/wastewater networks. Some of the related objectives may be to reduce the cascading of failures across these networks [Buldyrev et al., 2010], but also to develop an efficient repair strategy to restore services after disaster [Shekhtman et al., 2016]. Multilayer graphs can also be used to support robot indoor navigation in buildings [Ślusarczyk et al., 2017]. In this work, the graph represents a layout of the floors of the building with their interconnections. A layer is a floor containing rooms. An edge represents a direct transition between two rooms. Inter-layer edges model connections between floors. This kind of model reduces the amount of data to be analyzed by a robot.

The success of engineering projects depends on the capability to solve technological issues, as well as on the right management of social relationships among the stakeholders. Multilayer networks have been studied from this perspective too. Design Change Analysis is a typical example where multilayer networks have been used to trace how the social interactions between engineers and the evolution of technical requests influence the design of the final product [Pasqual and de Weck, 2012].

In the Architecture, Engineering, and Construction (AEC) domain, the visualization of interactions among the multiple stakeholders together with the dependencies among the components of the system to be built has also been studied for a long time. Although not explicitly described as multilayer networks, various techniques have been proposed to visually depict the different nature of nodes and edges. Howard and Petersen [2002] have used sociograms enhanced by the use of icons and various widths for edges in housing projects. Network sociograms have been adapted to replace the edge width by distance between vertices as an indicator of the intensity of the interaction [Thorpe and Mead, 2001]. Halin et al. [2004] have proposed modeling and visualizing these interactions as a hypergraph. The Interaction Visualization Framework (IVF) of Otjacques et al. [2006] proposes a way to characterize how such interactions can be visualized with 31 distinct features describing what is visualized, for whom and how. The vast number of instances of complex data sets produced across all these domains demands a visual approach to help understand it, and that approach will often be multilayer network visualization.

2.4 DATA SETS ACROSS DOMAINS

Just as Melançon [2006] noted that densities of standard networks vary across domains in the real world, it is important to realize that multilayer networks across domains may differ in many features. This is highlighted by Škrlj and Renoust [2020] in the strong differences of structure between social and biological multilayer networks.

The wide range of domains fortunately means that there exists a wide range of data sets that have been analyzed and visualized as multilayer networks, many of which are publicly available. Kivelä et al. [2014] provide a list of multiplex data sets covering a range of domains. Many data sets can also be found online. For example, the ComuneLab research unit for multilayer modeling and analysis of complex systems, at the Bruno Kessler Foundation Center for Information and Communication Technologies, provides an online repository of multiple types of multilayer network data sets,[2] all of which are publicly accessible and downloadable. The existence of such data sets makes it possible to test visualization techniques on data sets with the variety of properties that occur across domains.

2.5 CHAPTER SUMMARY

In this chapter, we described the wide range of application domains for the visual analytics of multilayer networks. It is evident that a multilayer network approach may be appropriate for those who are tasked with visualizing the complex systems within these domains. We have also seen that many fundamental concepts have existed for some time before the formalization of the concept of multilayer networks. Additionally, there are many related concepts that have been explored in the domain of network visualization, that may provide interesting research directions, and there are many sources of multilayer network data that can be used to support the exploration of new research on the topic. In the subsequent chapters, we will discuss the tasks that users of multilayer visualizations engage in, as well as the visualization and interaction

[2]https://comunelab.fbk.eu/data.php

approaches for layers and their constituent entities and associated attributes. We will provide examples from a range of the application domains discussed here as we do so.

CHAPTER 3

The Layer as an Entity

Contents

The notion of a layer is unique to multilayer networks, and at the center of the new challenges offered by multilayer network visualization. In this chapter, we discuss the notion of layers, and how it naturally appears from domain considerations. A specific focus on layers, considering them as entities of interest, allows a higher-level view into the structure of the complex system they belong to. We further discuss how layers can be integrated into data modeling, and how they are translated in terms of analysis and visualization.

3.1   MODELING LAYERS

The concept of layers is a natural notion that calls to intuition. Consider the modeling of a system from a given application domain. Most probably, the system being observed involves several types of nodes or links, that can be identified as such in the data (as seen in our examples in Chapter 1). For analytical reasons, it is worth making a distinction between nodes and edges of these different types when visualizing, manipulating, and reasoning about the data. Depending on the application domains and tasks, layers may serve different uses.

- When it is necessary to independently study different parts of the same system, keeping them separated in different layers.

- When it is necessary to combine entities: Layers can be combined or piled up.

- When in is necessary to compare different behaviors of the same group of entities of a system: Comparisons can be made across layers sharing the entities.

The notion of a layer often relates to that of scale or level: A layer gathers entities into a single, larger scale, or higher-level system. When defining what layers are, you decide on the granularity at which entities are considered.

Carefully defining what those fine grain layers are is thus critical when modeling a problem and organizing data. A layer will gather nodes and/or edges that can be characterized as homogeneously sharing one or more properties. A layer should rely on a relatively small number of these properties, and most of the time a single one. As a consequence, since nodes and edges will typically have multiple properties, they may be candidates to belong to multiple layers, which can be combined into (or derived from) aspects (see Section 3.1.3). Like any data definition, finding the right trade-off and defining the adequate number of layers or granularity, as well as defining aspects, is part of the modeling phase, and strongly depends on the domain application, data, and tasks.

3.1.1   LAYERS AS REAL ENTITIES OF THE NETWORK

Layers are much more than passive collections of network entities; they also interact together. Layers may overlap each other, and one may question how these are coupled together. For instance, one could observe different behaviors in the same social network (e.g., on Twitter, people mentioning, retweeting, and answering, as done by Omodei et al. [2015]). They may also transition one to another, for example the transition between a train network and a bus network [Gallotti and Barthelemy, 2015]. Note that even if a transition layer could be captured through a separated dedicated layer (as done in Skrlj et al. [2019]), it is obviously more explicit to have it as actual transitions between layers. Transitions may be ignored in some multiplex cases in which the assumption is that there is a transition between the same node belonging to multiple layers.

In a multilayer network, there are often multiple terminologies for common elements. Whatever criteria layers are defined upon, each layer will contain nodes sharing edges inside the layer, that we refer to as **intra-layer** edges, sometimes named *inner-layer* edges or *within-layer* edges. These are defined in contrast to **inter-layer** edges, *between-layer* edges, *transition* edges, or *coupling* edges, all referring to edges that connect a layer to another. The complex organization of layers is often investigated within one of these two families of edges: in multiplex networks, a complex structure of relationships between layers can be studied through layer overlap and correlation that are formed by **intra-layer** edges, as done by Nicosia and Latora [2015], Renoust et al. [2015], and Škrlj and Renoust [2020]; in more general networks with nodes of heterogeneous types, the structure of a network of layers appears thanks to **inter-layer** edges (e.g., De Domenico et al. [2013], Kivelä et al. [2014], and Mourchid et al. [2019]).

As a consequence, this organization of layers gives rise to different cases of multilayer networks, for which layers could be: overlapping, non overlapping, ordered or non ordered, have hierarchical relations together, or form a complex system with no obvious organization. The latter case can sometimes be captured in a network of layers, in which layers are the **entities**, i.e., nodes.

3.1.2   DEFINING LAYERS

Providing a precise definition of a layer also means providing a definition of a multilayer network as a whole since layers are *the* fundamental concept on which multilayer networks are built.

Recall from Chapter 1, a standard graph is often described by a tuple $G = (V, E)$ where V defines a set of vertices and E defines a set of edges (vertex pairs), such that $E \subseteq V \times V$. An intuitive definition of a multilayer network first consists in specifying which layers nodes belong to. Given a set of layers L, with an individual layer being defined as $l \in L$, and given that we we allow a node $v \in V$ to be part of some layers and not others, we may consider nodes in a multilayer graph as pairs $V_M \subseteq V \times L$. Edges $E_M \subseteq V_M \times V_M$ then connect pairs $(v, l), (v', l')$. An edge is said to be *intra* or *inter*-layer depending on whether $l = l'$ or $l \neq l'$. Edges connecting the same nodes across layers $v = v', l \neq l'$ are called *coupling* edges. If referring to the tensor formulation of multilayer networks, as done by De Domenico et al. [2013], each layer corresponds to a dimension in the tensor.

If the set of nodes is disjoint across layers, i.e., each layer has a different set of nodes that do not repeat, then the network has the same properties as a *node-colored* network. If networks have only inter-layer edges and no intra-layer edges, then we can consider them having the same properties as *bi-partite*, for two layers—or *k-partite* networks with more than two layers. Such graphs are found in text analytics, where layers coincide with types of entities, such as authors, books, publishers, and locations. Here, authors are connected to the books they wrote. The books are connected to publishers and publishers are connected to their locations. There is, however, a distinction to make, since *k-partite* may as well include undesired edges, as illustrated in Figure 3.1.

If only coupling edges exist across layers, then the multilayer network can be merged (see *Superposition* in Chapter 5) into one super-network visualization, with all nodes and multiple edges between nodes. These networks are called *edge-colored* or *multi-type*. A more detailed discussion on the relationship between multilayer networks and other network types can be found in Chapter 2.

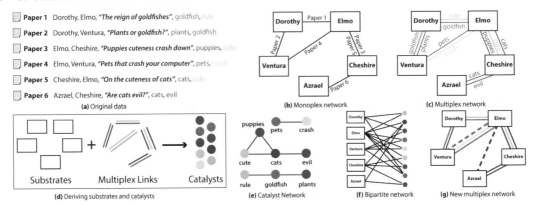

Figure 3.1: The definition of layers in *Detangler* [Renoust et al., 2015], here referred to as *catalysts*, from an attribute can seem equivalent to a *k-partite* definition. However, the construction rule combines a co-authoring aspect and a keyword aspects that is not equivalent to the *k*-partite graph.

3.1.3   ASPECTS

In their definition of multilayer networks, Kivelä et al. [2014] define what they call *aspects* as a way to characterize a set of layers relating to some problem-driven concepts. An aspect is a set of layers including the constituent nodes and edges based on the multilayer network topology and/or attributes. An example would be:

- Aspect A_1 capturing interaction between people in the context of their participation to events (e.g., conferences, Atzmueller et al. [2012]), with possible values *infovis* for interaction during the InfoVis conference, *eurovis* for interaction during the EuroVis conference, etc., generating the corresponding layers $l_{infovis}$ and $l_{eurovis}$;

- Aspect A_2 capturing co-authorship around themes (an example we borrow from Renoust et al. [2015]), with possible values k_i for co-authorship associated with some keyword k_i;

- Aspect A_3 capturing project partnership, with possible values p_j associated with specific program p_j, for example in Ghani et al. [2013];

- Aspects can also be used as an artefact to capture time or geographical position, and so forth.

A set of layers characterized by only a single aspect are referred to by Kivelä et al. [2014] as *elementary layers*. This terminology may be confusing, as an elementary layer from aspect A_x may contain more (or at least the same amount of) entities, i.e., nodes and edges, than a layer characterized by aspects A_x and A_y. Therefore, we prefer the term *single aspect layer*. A single aspect layer can only be part of one aspect. For example, the aspect $A_{conference} = (l_{infovis}, l_{eurovis}, l_{asonam})$ gathers interactions across three conferences (regardless of which year the conferences took place), while the single aspect layer $l_{infovis}$ describes all of the interactions during Infovis conferences. Any layer which is characterized by only the aspect $A_{conference}$, describing the interactions of people at a specific conference (regardless of which year), is a single aspect layer.

The fundamental difference between a single aspect layer and a layer is that a layer may be characterized by multiple aspects. Let us expand our example by adding another aspect characterizing the year in which interactions took place. With l_{YYYY} designating a layer formed of all interactions in any conference during year $YYYY$, $A_{years} = (l_{2019}, l_{2020}, l_{2021})$ is the aspect characterizing year of interaction. Each layer characterized by this aspect describes all of the interactions which took place in a specific year, regardless of the conference at which the interaction happened. One can define a set of layers characterized by both of these aspects to capture the interactions at specific conferences for specific years. This yields nine layers for our example, e.g., the layer $l_{infovis\_} \times \_2020$ is the intersection of the single aspect layer $l_{infovis}$, characterized by the aspect $A_{conference}$, and the single aspect layer l_{2020}, characterized by the aspect A_{years}. For another example of aspects, from the domain of life sciences, refer back to Chapter 1.

Depending on the application domain, aspects could also correspond to facets, dimensions, or attributes, or even to different families of ties such as Borgatti et al. [2009] define them. In practice, an aspect will characterize layers using a finite number of values. These can be obtained directly from the raw data, e.g., the values of a categorical attribute. They can also be obtained by transforming the data into a finite set of values, e.g., by binning continuous data. Further discussion specific to attributes of multilayer networks can be found in Chapter 7. Although aspects make useful tools to help define layers and understand the interaction of layers from a higher perspective, it is important to keep the resulting layers simple. Doing so helps avoid confusion and cognitive overload while handling analysis of the multilayer network (see Interdonato et al. [2020] for a survey of layer simplification techniques).

3.2   VISUALIZING AND MANIPULATING LAYERS

As already mentioned in Section 3.1, the notion of "layer" finds its strength in that it simultaneously resides in the domain and in the model. This makes its definition and operationalization intuitive, and *a priori* favors the affordance of any graphical representation of layers. Moreover, multilayer networks turn out to be flexible and allow building realistic models, since layers can capture any type of nodes and edges.

As a consequence, a multilayer network may contain far more layers than what is needed when investigating a specific question. In this context, manipulating layers in different ways may be necessary before further investigating, or processing any data on nodes and/or edges. Layers thus require the design of specific approaches allowing users to conduct such manipulations, as described by Interdonato et al. [2020]. We further dive into layer related tasks in Chapter 4.

3.2.1   LAYER ANALYSIS AND VISUALIZATION

While the analysis of multilayer networks is an active field of research, plenty of analytical tools are already available. Many tools are designed to identify the nature of elements in their larger multilayer graph context, i.e., nodes and edges of interest within layers. Others tools contrast in being specifically designed to measure layers, and compare between them, i.e., considering layers as *entities*. An exhaustive list is beyond the scope of this book, but one can refer to Aleta and Moreno [2019], Battiston et al. [2014], Bródka et al. [2018], Pamfil et al. [2020], or Renoust et al. [2014] among others; readily available packages such as *MuxViz*[1] De Domenico et al. [2015], *Multinet*[2] implementing the work of Dickison et al. [2016b], and *Py3Plex*[3] [Skrlj et al., 2019] also includes some layer specific measures. For a more extensive discussion on multilayer networks analytics—not limited to layers, we invite the interested reader to further dive into Chapter 6.

Considering layers as entities enables their abstraction into visual entities, helping to reduce visual complexity. Doing so may help give a structural overview of the multilayer network. This is achieved in *Detangler* [Renoust et al., 2015] where a network of layers represents the whole multilayer network (see the *catalyst network* in Figure 3.1). In *MuxViz* [De Domenico et al., 2015], the annular visualization (described in Chapter 7) and the layer correlation matrices also represent layer entities with their properties and similarities.

In later chapters we look at visualizing the constituent entities of layers in more detail. Chapter 5 examines the visualization of nodes and relationships across layers. Chapter 7 focuses on the visualization of multilayer network attribute data.

Since layers can be abstracted as entities and compared, the discussion naturally strays toward the different operations on layers. The work of Interdonato et al. [2020] provides good examples of operations on layers, but the integration of such manipulations always needs to be put into perspective with the user's intent and tasks. We discuss tasks for multilayer networks in Chapter 4.

Implementing layer manipulations requires interaction design. Layers can be seen as ordered entities. Their manipulation helps derive different network visualization as in the *Donatien* application of Hascoët and Dragicevic [2012]. Layer entities may form a high-level network. Interacting with this high-level network guides the linked highlighting lasso selection of *Detangler* [Renoust et al., 2015]. This latter work especially addresses the issue of interaction design for layers as entities, unique to multilayer graphs. The authors first define mapping functions between multilayer nodes (they call *substrates*) and layer entities (they call *catalysts*) based on multilayer subgraphs (Figure 3.2). These are embedded as a linked highlighting lasso selection. The different mapping functions exist both ways and are piped together to form what they refer to as a *leapfrog interaction* upon double-click on the selection lasso brush (Figure 3.3). Further

---

[1]http://muxviz.net
[2]https://cran.r-project.org/package=multinet
[3]https://pypi.org/project/py3plex/

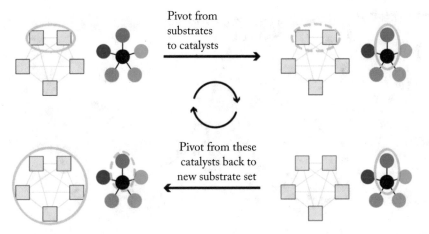

Figure 3.2: *Detangler* [Renoust et al., 2015] sets multiple mapping functions between nodes (*substrates*) and layers (*catalysts*) of their multilayer network that allows coordination of their selection.

details on interaction techniques for multilayer networks can be found in Chapter 6. *Detangler* is the only approach that we have identified that further implements a layer analysis component (using layer entanglement). The measure supports the interactions, augmenting the lasso interaction with colors, and directly mapped to layer entity size (as illustrated in Figure 3.3).

3.2.2 MODELS OF LAYERS

There is a variety of data definitions found in the visualization literature on which visual representations of networks with multilayer characteristics are built. We may distinguish three major definitions: layers inferred from data attributes, layers formed from heterogeneous sources/aggregated data, and layers which correspond to different types of relationships.

Only a few approaches explicitly mention the use of multilayer networks (both as data underlying the visualization and as a visual encoding). Most systems dealing with multivariate networks couple relational data with node and edge attributes, e.g., Bezerianos et al. [2010], Bigelow et al. [2019], Heer and Perer [2014], Shen et al. [2006], and Wattenberg [2006] often using table-based representations, e.g., Heer and Perer [2014] and Kerren and Schreiber [2014]; they do not consider any data or attribute specifying a layer structure. Cao et al. [2010] consider *classes* of entities they call *"facet"* which appear to naturally map to layers of nodes (see Section 2.2.2). *Origraph*, from Bigelow et al. [2019], is an interactive system. It allows the creation of networks inferred from a data set with a meta-model definition, which includes potential multiple definitions of edges. Among all, the work of Pretorius and Van Wijk [2008] is a notable exception. It introduces the notion of layers without using the term, and explicitly defines nodes as Cartesian products of attributes.

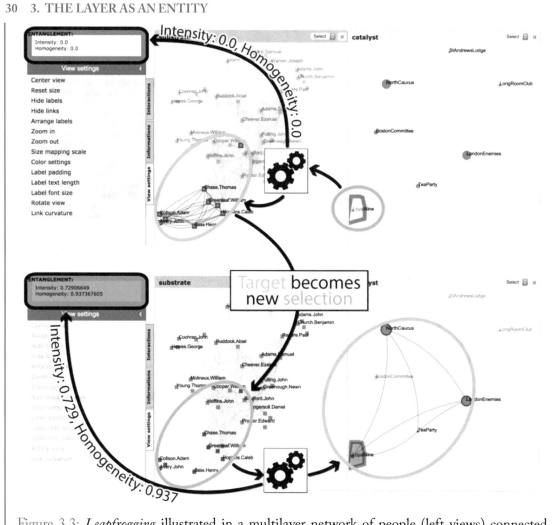

Figure 3.3: *Leapfrogging* illustrated in a multilayer network of people (left views) connected through multiple social circles (right views). It allows users to realize two tasks in a single operation. It starts with a cross layer entity connectivity task: a lasso selection is made from a social circle on the top right (*LoyalNine*); highlighted with its corresponding social network on the top left. The leapfrog allow either a cross-layer entity comparison when the selection started with node entities, or a topological layer comparison task, when the selection started with layer entities. When leapfrogging is triggered the highlighted social network becomes a new source selection (bottom left), and a network of social circles connected through these people is now linked highlighted (bottom right, with *NorthCaucus*, *LondonEnemies*, and *TeaParty* highlighted). Multilayer network measures enrich the visual encoding: the lasso colors reflect the selection's homogeneity and intensity. Highlighted node size correspond to influence. It suggest that the *LoyalNine* have their lowest influence on the *TeaParty*.

Other systems and approaches infer multilayer structure by aggregating data from multiple sources, whether from databases [Kohlbacher et al., 2014] or from a collection of ego networks (as in Dunbar et al. [2015]) and/or personal data [Huang et al., 2015]. Interestingly enough, some systems do not directly target the visualization of multilayer networks, but use multiplex and/or hypergraph representations to build query graphs or summarize query responses (Ren et al. [2018b], Renoust et al. [2020], Shadoan and Weaver [2013], and Tu and Shen [2013]).

Coming from the domain of complex systems, *MuxViz* [De Domenico et al., 2015] relies on the exact definition and implementation introduced by Kivelä et al. [2014]. Authors mentioning explicit use of *MuxViz*, e.g., Gallotti and Barthelemy [2015], also use the same definition. Layers originating from aspects of the network, such as time or node/edge type, are quite similar to the facets described in Hadlak et al. [2015]. *Detangler* [Renoust et al., 2015] relies on an explicit encoding of layers, with a goal of allowing an easy exploration of inter-layer correlation. Making a distinction between layers as being either structural or functional (or of any other type) may be useful depending on the pursued goal [Agarwal et al., 2017].

3.2.3   LAYER COMPARISON AND OVERVIEW

For some graph comparison tasks, a detailed inspection of local structure is not required and only an overview is necessary. Here, a summary approach not showing individual nodes or edges can be taken, similar to the layer entity network introduced previously in *Detangler*. In the *Graph Thumbnail* representation of Yoghourdjian et al. [2018], a graph is decomposed hierarchically using what the authors refer to as a K-core component clustering, or KC3, decomposition. In this decomposition, the top two layers of the hierarchy are defined by the single and bi-connected components of the graph, respectively. The third level consists of three cores within a bi-connected component. The subsequent levels are defined by k-cores (where $k \leq 3$). As seen in Figure 3.4, this hierarchical decomposition is visualized using a hierarchy of circles, positioned using a circle packing algorithm. It is adorned with further node and edge distribution information. This type of visualization reveals structural information about a network, allowing for rapid comparison of networks. It can be easily applied to the comparison of layers in a multilayer network (supporting Task category **D2** of Chapter 4).

Other existing systems also provide the ability to view graph structure information as a form of summary visualization. *ManyNets* [Freire et al., 2010] is an approach that uses simple attribute based visualizations, such as bar charts and histograms, as a means of summarizing and comparing networks, as seen in Figure 3.5. The simple charts show metrics that describe the structure of the graph. The set of charts describing a network is referred to as a "network-fingerprint" and the tabular presentation allows for easy comparison and sorting across networks (or layers, depending on the nomenclature chosen). The annular visualization of *MuxViz* [De Domenico et al., 2015], also allows for comparison of structure across layers (see Chapter 7). In general, structural visualizations are used to compare graphs, and hence can be used to com-

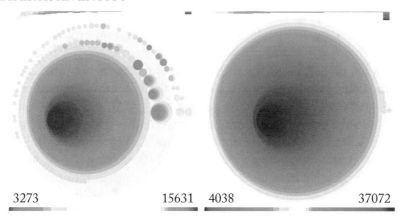

Figure 3.4: Graph thumbnails generated to compare the structure of two graphs for the analysis of protein interaction networks, taken from Yoghourdjian et al. [2018]. The color and size of the circles represent structural information in a hierarchical decomposition of the graphs.

ID	TRUST	Edge count	Edge-vertex ratio	Vertex count	Edge density	In degree	Out degree	large ▾ and dense	
ego-98		137	7.211	19	0.401				
ego-176		80	4.211	19	0.231				
ego-41		86	5.059	17	0.316				
ego-371		78	5.571	14	0.429				
ego-363		27	4.5	6	0.9				
ego-156		64	4	16	0.262				
ego-368		67	5.154	13	0.429				
ego-8		66	5.077	13	0.423				
ego-89		18	3.6	5	0.9				

filmtrust/1 nets - all radius 1.5 ego nets - 'result = Column['Vertex count'] > 4' - 'result = Column['Vertex count'] < 20'

Figure 3.5: The list view of the *ManyNets* application [Freire et al., 2010], summarizing attributes of networks using bar charts. The vertical barcharts show the distribution of attribute values and the green and red stacked horizontal bar is a combined score based on several inputs.

pare layers (see Task categories **D1** and **D2** of Chapter 4). For a complete taxonomy of visual comparison in information visualization see the work of Gleicher et al. [2011].

3.3   CHALLENGES AND OPPORTUNITIES

Many of the approaches to defining layers, particularly from the information visualization domain, do not explicitly mention that the data is a multilayer network. An important part of understanding the data is determining what aspects (and hence layers) need to be visualized to support the users goals as early in the design process as possible. As described in Chapter 1, layers can be considered a characteristic of the multilayer system as a whole, defined either by a physical reality or the system being modeled. However, there are still multiple ways to determine the set of layers for analysis.

3.3.1   MODELING OF REAL-WORLD CONCEPTS FROM THE DATA

Models of real systems often begin with raw data and not a graph. However, in much of the work we have examined, systems are presented with fully organized and cleaned data sets, e.g., Kairam et al. [2015] and Shi et al. [2014]. Within an application domain, generating a multilayer data set for analysis is often by itself a significant focus of the work (e.g., Ducruet [2017], Gallotti and Barthelemy [2015], and Zeng and Battiston [2016]) which is independent of visualization. It is already recognized that creating a general purpose graph from real data is a challenge [Kandel et al., 2011, Srinivasan et al., 2018], and doing so across multiple layers can be considered even more challenging. Existing approaches, such as those of Bigelow et al. [2019], Heer and Perer [2014], and Srinivasan et al. [2018], consider the problem from a general graph point of view and could be developed further to consider aspect and layer definition.

3.3.2   CREATIVE DEFINITION OF LAYERS

When modeling layers it is easy to consider a node type attribute to characterize a single aspect and encode data into layers. However, it is worth emphasizing that there are many other options. Multiple aspects can be, and often are, combined together to define layers, providing the ability to slice the data in ways that may be more useful to an analyst. For example, in the biological domain, similar to our example from Chapter 1, one aspect could be omics level, while another aspect could be species, resulting in layers that describe an omics level for a specific species. Edge types are also used in many cases to generate layers (usually in multiplex cases such as Ducruet [2017] and Renoust et al. [2015]). It is also possible for layers to be defined based on characteristics that are not directly encoded, e.g., a set of layers where each layer is the ego-networks of a specific individual in a social network. The full range of approaches to layer definition is yet to be fully exploited in the visualization domain, providing an opportunity for new and interesting research. It is worth remembering the advice of Kivelä et al. [2014], and be "creative."

Additionally, when layers are defined manually by domain experts, the layers might be purely subjective, based on the experts' intuition. Objective reasons determining the definition of these specific layers might be latent from a combination of attributes. Inferring the aspects which gave rise to such layers, and how they were combined, clearly is an open challenge.

3.3.3   ANALYTICAL GENERATION OF LAYERS

The raw data may not map to the real world concepts embodied in a system and may require some processing. If layers are not immediately forthcoming, transformations may be applied to the data. Interdonato et al. [2020] describe such transformation as part of a set of different techniques to simplify a multilayer network for analysis. These techniques include projection and graph embedding, which essentially is dimensionality reduction applied to the graph data. Alternatively, a clustering approach might reveal structure not explicitly encoded in the data. Consider the example of a predator-prey network. A topological clustering may group animals based on geography, even if geography is not explicitly encoded in the data. While the process is analytical, it still results in a layering that reflects the reality of the system being modeled. Degree-of-Interest (DOI) functions suggest nodes for inclusion based on what the user has already characterized as interesting. This approach has already been used by the *Refinery* application of Kairam et al. [2015] and may also be applied to data sets that are explicitly multilayer, as seen in Laumond et al. [2019].

Generating layers analytically may cover a larger spectrum than required for the domain analysis. As a consequence, a lot of noise may be included in the resulting multilayer network. A wide range of techniques to simplify such a multilayer network for analysis described by Interdonato et al. [2020] may be useful in this case.

Such analytic techniques surely open ways for new and novel approaches to layer creation. Projection and other techniques for transforming layers are also discussed in Chapters 4 and 6.

3.3.4   SUMMARY VISUAL COMPARISON ACROSS LAYERS

We have discussed summary visualization of layers and seen that there are existing network summary techniques. They can be adapted to summarize the layers of a multilayer network, but as-is only when considering each layer as a stand alone entity. There is no visualization of the edges between layers, or comparative visualization of how data changes across layers. Tools such as *Detangler* from Renoust et al. [2015] and *MuxViz* from De Domenico et al. [2015] provide some functionality for understanding the relationships between layers. However, the development of summarization techniques for multilayer networks that take into the account the relationships and differences between layers is a research challenge that has yet to be fully addressed. The alluvial diagrams used by Rosvall and Bergstrom [2010] (see Figure 3.6) to visualize changes in temporal networks may be an inspiration for visualizing differences across layers which are linked by many coupling edges.

3.3.5   THE ASPECT AS AN ENTITY

We have seen that handling layers as entities can provide a higher-level view on a complex system. Doing so can help drive a visual analytics tool such as *Detangler*, from Renoust et al. [2015]. There is an open avenue of research for extending this concept, and analyzing and visualizing aspects in multilayer networks. Of course, the discussion on aspects may intersect with

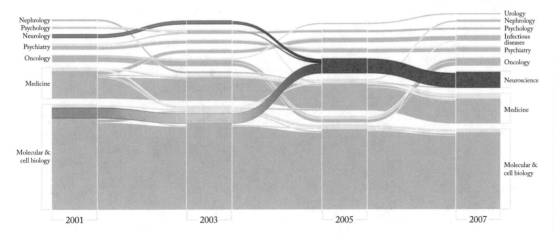

Figure 3.6: An alluvial diagram from Rosvall and Bergstrom [2010] captures changes in a temporal network (here clustering in a citation network). Each block in a column represents a field (cluster) and size reflect their citations. The field neurosience is selected, and darker colors indicate the significance of the subsets of each cluster.

attributes visualization (detailed in Chapter 7). However, taking the visualization of multilayer networks to the level of its aspects while handling them as entities remains a completely open challenge. Considering *MultiAspect networks* as defined by Wehmuth et al. [2016, 2017] may be an interesting step toward that direction.

CHAPTER 4

Task Taxonomy for Multilayer Networks

Contents

There are visualization tasks that make multilayer network visualization unique. In the definition of multilayer networks, layers are entities (see Chapters 1 and 3) that are fully part of the network structure. As a result, layer-related tasks are not considered abstract or high level. They are as fundamental a part of a graph task taxonomy as nodes and edges. Although most task taxonomies that have been developed in the domain of network visualization so far (e.g., Ahn et al. [2013], Beck et al. [2017], Kerracher et al. [2015], Lee et al. [2006], Nobre et al. [2019], and Pretorius et al. [2014]) do not directly address multilayer networks *per se*, they could be further adapted or extended to target multilayer network visualization. For instance, layers may be leveraged when performing a search task, specifying the searched element(s) as belonging to a given layer. Conversely, the answer to a search query may take advantage of the network's layer structure to organize its output, sorting queried elements according to the layers they belong to.

4.1 TAXONOMY OF MULTILAYER NETWORK TASKS

We present below a revised and extended version of the taxonomy for multilayer network tasks initially proposed in McGee et al. [2019a]. We report on different approaches or systems supporting tasks relevant to multilayer networks. These are summarized in Table 4.1 with a non-exhaustive selection of systems and techniques. In many cases, authors have not explicitly expressed tasks in terms of layers, but rather referring to properties of the data they consider. For

Table 4.1: A non-exhaustive selection of techniques/systems relevant to multilayer networks, that they either implicitly or explicitly support. Notes in parentheses refer to task labeling/naming as indicated by authors in their paper.

A—Cross-layer connectivity

Collins and Carpendale [2007], Santamaría et al. [2008], Ghani et al. [2013] (Q1b,c),
Kairam et al. [2015] (assoc. browsing), Renoust et al. [2015] (leap-frogging),
De Domenico et al. [2015], Bourqui et al. [2016], Liu et al. [2017] (Q4)

B—Cross-layer entity comparison

Santamaría et al. [2008], Holten and Van Wijk [2008], Bezerianos et al. [2010] (multi-facet query),
Cao et al. [2010], Freire et al. [2010], Dunne et al. [2012], Ghani et al. [2013] (Q1a),
De Domenico et al. [2015], Xia et al. [2015], Renoust et al. [2015] (also leap-frogging),
Bourqui et al. [2016]

C1—Layer reconfiguration

Shen et al. [2006], Krzywinski et al. [2011], Hascoët and Dragicevic [2012], Ghani et al. [2013],
Cao et al. [2015], Xia et al. [2015], Vehlow et al. [2015], Liu et al. [2017] (Q1, Q2)

C2—Layer derivation

Shen et al. [2006], Krzywinski et al. [2011], Renoust et al. [2015], Melançon et al. [2020],
Cottica et al. [2020]

D1—Numerical comparison

Shen et al. [2006], Stasko et al. [2008] (disparity), Krzywinski et al. [2009], Krzywinski et al. [2011],
De Domenico et al. [2015], Renoust et al. [2015], Cao et al. [2015], Liu et al. [2017] (Q3)

D2—Topological comparison

Shen et al. [2006], Wattenberg [2006] (roll-up), Stasko et al. [2008] (disparity), Santamaría et al. [2008],
Holten and Van Wijk [2008], Bezerianos et al. [2010] (R5, R12), Cao et al. [2010],
Hascoët and Dragicevic [2012], Ghani et al. [2013] (Q2c), Renoust et al. [2015], Kairam et al. [2015],
Vehlow et al. [2015], De Domenico et al. [2015], Cao et al. [2015], Liu et al. [2017] (Q3)

---

example, this is the case for authors considering tasks related to group comparison or reconfiguration, such as Cao et al. [2015] and Hascoët and Dragicevic [2012]. To address this situation, we propose task categories specific to multilayer networks. We target tasks directly involving visualization, as opposed to tasks that can be addressed through computational means only.

Tasks specific to multilayer networks revolve around the notion of a layer. Tasks often boil down to handling elements within one layer, or across several layers, or directly the layers themselves. These often lead to higher level tasks, which are also critical for visual analytics (e.g., understanding group interaction or communication patterns in social networks). Some tasks may involve a temporal dimension as well, such as tracking the evolution of nodes or edges at different moments.

In the survey work of Pretorius et al. [2014], a task is schematized as a process: *Select entity → Select property → Perform analytic activity*. In this case an entity refers to a node or an edge. We see here an important difference with the process of performing a task on a multi-layer network involving layers. Conceptually speaking, layers are genuine building blocks of a multilayer network. They are neither a simple (sub-)network nor a mere property of a node or edge. They are a conceptual construct that fully enters the analytic process when performing a task (involving the multilayer nature of the network). Thus, we describe below more than just yet another visualization task taxonomy.

4.1.1   TASK CATEGORY A – *ENTITY CONNECTIVITY ACROSS LAYERS*

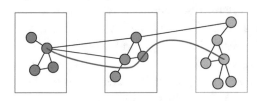

Tasks in this category aim at exploring and/or inspecting connectivity involving paths traversing multiple layers (i.e., inter-layer paths). Understanding how shortest paths expand across layers, e.g., Ghariblou et al. [2017], and inspecting what nodes do occur on these paths are typical examples of tasks in this category. Being able to explore cross layer connectivity has been identified as an important user task [Ghani et al., 2013]. Associative browsing in *Refinery* [Kairam et al., 2015] is a good illustration of a cross-layer connectivity task. It performs cross-layer random walks and collects nodes from different layers in a single view. The leap-frogging operation in Detangler [Renoust et al., 2015] is another good illustration of cross layer connectivity building a dual view, reflecting how/what layers get involved when hopping from node to node (see also Section 6.1).

4.1.2   TASK CATEGORY B – *ENTITY COMPARISON ACROSS LAYERS*

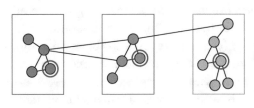

Tasks in this category aim at comparing entities (nodes and/or edges) across different layers. Comparing attributes or metric values of a node (or a subset of nodes) across layers, and comparing the neighborhood of a node across layers are examples of such tasks. This requires the ability to query entities across layers. The task may concern the same (set of) node(s) over several layers (e.g., are my friends also my colleagues in an online social network platform?); or distinct nodes that are somehow linked across different layers. *Jigsaw*, by Stasko et al. [2008], typically supports this task category by allowing users to identify entities (persons, places, etc.) through several documents (seen as layers in a multilayer document network). Other good examples are *FacetAtlas'* [Cao et al., 2010] multi-facet query box, or the animated drag & drop of Renoust et al. [2019b].

4.1.3   TASK CATEGORY C – *LAYER STRUCTURE MANIPULATIONS*

Tasks in this category aim at manipulating the layer structure itself. Such manipulations may allow for previously unseen relationships and structure to be revealed, and allow for new perspectives on the underlying data. This task category also concerns manipulations of aspects which characterize collections of layers (see Section 3.1.3).

Task Category C1 – *Reconfiguration: Filter, Select, Aggregate, Split, Merge*

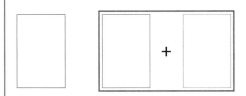

Tasks in this sub-category aim at modifying the layer structure. No new content within layers is created. Filtering, selecting, or changing the granularity of visualization elements on existing layers are typical tasks. This last task is also called aggregate/segregate in Brehmer and Munzner [2013]. Combining layers through drag & drop operations, as in the work of Hascoët and Dragicevic [2012], is a perfect illustration; another example is *g-Miner* [Cao et al., 2015] which allows to create, edit, or refine the grouping of elements.

Task Category C2 – *Derive Layers: Project, Transform*

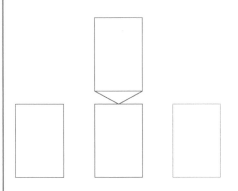

Tasks in this sub-category aim at creating new layers by projection or transformation of existing layers. Data comprised in a layer (or several layers) is somehow modified. Inducing a layer (from pre-existing layers for instance) to create a new layer comprising links that were not already present is a good example of a derivation. Projections typically take place when layers originate from a bipartite graph (recall Chapter 2); in this case, the projection onto a single modality (i.e., node type) of the bipartite graph, with the other modality serving as a basis for a layer or multiple layers, may be seen as inducing inter-layer edges (see Melançon et al. [2020], Renoust et al. [2015]). Cottica et al. [2020] uses more elaborate transforms where a semantic layer and a social network layer are visualized using multiple coordinated views constructed from patterns simultaneously involving actors, messages, and ethnographic codes.

4.1.4 TASK CATEGORY D – *LAYER COMPARISONS*

Tasks in this category support comparing layers to one another based on numerical measures or topological features. This task category also concerns comparison of aspects (see Section 3.1.3).

Task Category D1 – *Comparison Based on Numerical Attributes*

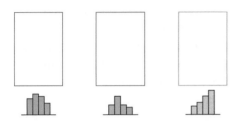

Tasks in this category support comparing layers to one another based on numerical measures summarizing layer content and structure. Typically, layers could be compared by looking at how node degree distributions compare layer-wise. *OntoVis* (Shen et al. [2006], where layers map to node types) supports layer comparison tasks using a metric they call (inter-layer) *node disparity*. Pretorius and Van Wijk [2008] propose a quite elaborate approach and system to perform multi-attribute based layer comparison.

Task Category D2 – *Comparison Based on Topological/Connectivity Patterns*

Tasks in this category support comparing layers through non-numerical topological features of layers (e.g., group structure). A layer could be hierarchical (inheritance), while another could show a strong scale-free structure, for instance. The work by Vehlow et al. [2015] is a typical technique allowing to compare group structure across layers. The design rationales described in the work of Bezerianos et al. [2010] concerning the users' need for attribute visualization (referred to as R5 by the authors) and consensus/similarity (referred to as R12) across clusters are good illustrations of this type of task. In the biological domain, the tool *Netaligner*, from Pache et al. [2012], compares different biological networks, which can be considered analogous to layers, and visualizes them in a single visualization to allow a user to determine how well the networks are aligned, i.e., overlapping in terms of nodes and edges. Leapfrogging from the layer graph in *Detangler* specifically performs a layer comparison through their shared nodes.

4.2 RELATED TASKS TAXONOMIES IN VISUALIZATION

Numerous literature surveys (e.g., Ahn et al. [2013], Beck et al. [2017], Kerracher et al. [2015], Lee et al. [2006], Nobre et al. [2019], and Pretorius et al. [2014]) list tasks relevant to the visual analysis of different types of networks (general, evolving, multivariate, etc.) and tasks have been proposed on a domain specific basis, e.g., Kohlbacher et al. [2014] and Murray et al. [2017].

Lee et al. [2006] provide a general graph task taxonomy. At its top level it considers *Topology Based Tasks*, *Attribute Based Tasks*, *Browsing Tasks*, and *Overview Tasks*. It explicitly specifies that the high level tasks of comparison of graphs and identifying graph changes over time are not covered by the taxonomy. Pretorius et al. [2014] focus on multivariate networks. The highest level of their taxonomy divides tasks as follows: *Structure-Based Tasks*, *Attribute-Based Tasks*, *Browsing Tasks*, and *Estimation Tasks*. The category *Estimation Tasks* is further subdivided and more detailed than the *Overview Tasks* of Lee et al. [2006]. The name was chosen to capture that these tasks are not easily definable using lower-level tasks and are considered more high level, and are not focused on giving precise answers. Within this categorization there is a comparison task, which may be of some relevance for multilayer graphs. It covers comparing information at different stages of a network's development, and determining causation, i.e., providing an explanation for the differences between two snapshots of a changing network.

While Pretorius et al. [2014] do consider graph changes as part of their multivariate network task taxonomy, the taxonomies of Kerracher et al. [2015] and Ahn et al. [2013] both focus specifically on dynamic networks, also known as evolving or temporal networks. At the highest level, Ahn et al. [2013]'s taxonomy focuses on three groupings: *Entities*, *Properties* and *Temporal Features*. The temporal features are grouped as *Individual Events*, the *Shape of Change*, and the *Rate of Change*. These are considered from the individual entity level to the entire network level, and for both structural and domain properties. Kerracher et al. [2015]'s taxonomy builds on the non-network specific taxonomy of Andrienko and Andrienko [2006] by extending it to include network data. It considers both elementary and synoptic tasks, as defined by Andrienko and Andrienko [2006]. Elementary tasks involve individual items and characteristics, synoptic tasks involve sets of items considered as entities. However, Kerracher et al. [2015] further divides synoptic tasks into three categories. These are tasks considering graph subsets, tasks considering temporal subsets, and tasks considering both graph and temporal subsets. The taxonomy differs from Ahn et al. [2013]'s in that it focuses more on the tasks that data items take part in, rather than the data items themselves. It also considers a more general concept of pattern changes that captures relational change in the network, as well as considering tasks which provide context for graph evolution.

Most of the previous cited taxonomies focus on low level tasks, i.e., tasks focusing on individual entities such as nodes and edges independent of user context, motivation or background knowledge. A higher-level domain-specific approach can be seen in the work of Murray et al. [2017]. This taxonomy proposes, in the context of biological pathway visualization, tasks concerning comparison, attribute analysis, and annotation that relate to multilayer networks. Multilayer network tasks require a low-level approach as the layer itself is a low level entity. However, the wide range of application domains, as discussed in Chapter 2, also requires a more abstract definition of tasks.

Brehmer and Munzner [2013] address the gap between many high- and low-level task taxonomies by proposing a typology which breaks down a task into several parts spanning three dimensions: *why*, *how*, *what*. Hence, this typology provides a higher level reading of our taxonomy. As we consider layer-centered tasks, this lays down a common ground to all task categories. The *why* part of the task specifies the motivation to perform a task and, in a sense, is not data-dependent. However, layers clearly act on the *what* and *how* dimensions:

- Tasks in categories **A** and **B** can be seen as particular instances of *locate* or *explore* tasks, where layers somehow structure the way elements are searched.

- Tasks in categories **C1** and **C2** may be seen as belonging to *explore* tasks or, to some extent, to *summarize* tasks. Indeed, rearranging layers could well be done tentatively during early exploration, or in confirmatory mode in order to organize data in view of hypothesis testing, or following (layer) correlation analysis.

- Tasks in categories **D1** and **D2** clearly are *compare* tasks where layers appear as input ingredient to the *what* dimension, and where the output ingredients vary according to the type of information that is sought.

Recently, Interdonato et al. [2020] provided a taxonomy of techniques that focuses on one specific goal, multilayer network simplification. The complexity of multilayer networks can make them difficult to analyze. Simplification techniques aim to aid an analyst by removing irrelevant information and making the size of the network more manageable via selection aggregation and transformation based techniques. These simplification tasks fall under Task category **C** in our taxonomy. *Selection*, *filtering*, *sampling*, and *grouping/flattening* fall under Task category **C1** and *transformation*, and *compression/summarization* fall under **C2**.

4.3 CHALLENGES AND OPPORTUNITIES

By considering layers as low-level elements, we have built a taxonomy for low-level tasks. Other taxonomy approaches may be considered, and new tasks may become more apparent as multilayer network visualization increases in popularity.

4.3.1 TASKS FOR DEFINING AN INITIAL LAYERING

While some application domains inherently provide layer definitions (for example in biology), defining an initial layer structure from a flat network is an important task which may be considered as a category in itself. Our task taxonomy does not explicitly define this initial layering as it is a taxonomy of task for multilayer networks, i.e., those which already contain layers as entities. However, as a starting point for future research, if we consider the initial network as a multilayer network with only a single layer, then Task categories **C1** and **C2** reflect what may be done as part of this initial layering.

4.3.2   CONSIDERATION OF OTHER TAXONOMIES

Our approach focuses on the layer as a network level entity. Other taxonomies focus on domains, or in the case of Brehmer and Munzner [2013] on abstract tasks, discussing the *how* and the *why* as well as the *what* of a task. In the future, it may be beneficial to also consider layers in the context of these taxonomies. As discussed in Chapter 2, taxonomies from other related fields such as dynamic graph visualization [Beck et al., 2017] or faceted visualization [Hadlak et al., 2015] may provide some further inspiration when it comes to exploring the task space for multilayer networks.

CHAPTER 5

Visualization of Nodes and Relationships Across Layers

Contents

In this chapter, we examine visualizing multilayer networks and the constituents of their layers in detail. Abstraction of layers and the visualization of layers as entities are discussed in Chapter 3. From a multilayer network perspective, previous work in network visualization techniques may be classified based on their awareness of the notion of a layer and how it impacts the presentation of the nodes and their relationships. When this is the case, multilayer networks are visually encoded in a spatial representation; the networks are also interacted with and manipulated as visual objects in their own right, as detailed in Chapter 6.

The visual representation of a multilayer network needs to encode the layers, their constituent nodes, the inter-layer edges (those edges between the layers, including the so-called coupling edges), the intra-layer edges (those edges within a layer) as well as their possible attributes. This chapter does not consider attributes, these are discussed in Chapter 7.

1D layer   2D layer   3D layer

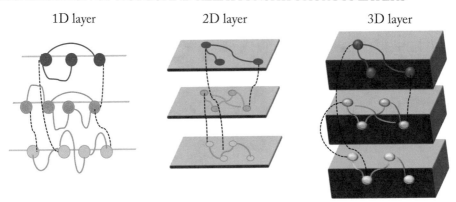

Figure 5.1: Dimensionality of layer visualizations.

The visual design of multilayer networks differentiates between matrix-based and node-link based representations. Although we discuss matrix-based visualizations at the beginning of the chapter, the primary focus is on node-link visualizations. We consider two main visual encodings of layers and nodes/edges in node-link representations: (1) position, the strongest of all perceptual cues (in the sense of the ranking of perceptual tasks by Mackinlay [1986]), is frequently used to distinguish which layer a node belongs to; and (2) color hue is also a strong cue for nominal data per Mackinlay [1986] and, as would be expected, is frequently used to encode layer membership for both nodes and edges (Bourqui et al. [2016], De Domenico et al. [2015], and Fung et al. [2009]). Color hue is used most prominently when multiple layers are superimposed on top of each other (as done by Ducruet [2017]). It can be also used to enhance the juxtaposed representations (as can be seen in Figure 5.10). Shape and texture are other prominent nominal cues mentioned by Mackinlay [1986], however they are not frequently used to distinguish between layers in multilayer applications. Weaker cues such as size and saturation are not used for indicating layer membership for nodes and edges, and are usually reserved for representing quantitative or ordinal attribute values associated with a node or edge.

The position of nodes and edges within a layer determines the dimensionality of the visual encoding. For example, the nodes that belong to a layer can be positioned along a line, thus creating a 1-dimensional layer representation. Alternatively, they can be positioned in a rectangular or cubic space, i.e., 2D or 3D space (see Figure 5.1 for an illustration). In addition to this basic dimensionality of inter-layer edges and associated nodes, the position of layers relative to each other, and the shape of the inter- and intra-layer edges play an important role in the visual design.

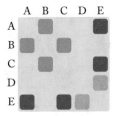 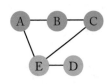

Figure 5.2: A simple illustrative example of a matrix visualization and the corresponding node-link visualization. Color is used to distinguish between layers (purple and orange). However, in this example, for the inter-layer edges in the matrix view, a separate color (black in this case) is required for visual encoding. This helps ensure that these edges are not misunderstood as being within a layer.

5.1   MATRIX-BASED VISUALIZATIONS

Matrix-based visualizations consist of laying out nodes as the rows and columns of a two-way table. A link between two nodes is often represented as a square at the intersection of the associated row and column (Figure 5.2). Standard node-link representations of graphs give equal importance to nodes and links and aim usually to convey structural properties of the graph at hand. They may however be difficult to read, due to edge clutter, particularly as graph size increases, and for the more complex networks encountered in many real usage scenarios. When dealing with large and/or dense graphs, matrix-based representations were found to be more readable than node-link diagrams for many tasks, per Ghoniem et al. [2005b]—path finding the notable and frequently cited exception. Matrix visualizations have also been shown to outperform node-link visualizations for some tasks involving clusters of nodes [Okoe et al., 2018]. Color is often used to encode the weight of the links, when link attribute values are available. This makes matrices very similar, if not identical in essence, to heatmap views frequently used in biology and other domains [Wilkinson and Friendly, 2009]. Other visual designs include using circles at the intersection of rows and columns with size and color encoding link attribute values [Chuang et al., 2012]. Matrix representations have been used to visualize homogeneous graphs (nodes of one type), e.g., in software engineering [Van Ham, 2003], and bipartite (or 2-mode) graphs, e.g., in software performance tuning; see Ghoniem et al. [2005a]. In the work by Ghoniem et al. [2005a], the analyst needed to compare the graph structure and edge weight distribution before/after a software revision by contrasting the corresponding matrix representations. They also monitored the runtime behavior of the software through the animation of the matrix representation showing the graph structure and edge weights changing over time.

The ability to detect link patterns in a matrix view is conditioned by the use of an appropriate ordering of rows and columns. Various seriation algorithms (Chen [2002], Fekete [2015], and Fung et al. [2009]) reorder the rows and columns of the matrix to create dense rectangular blocks of links. Often clustering algorithms are used to guide matrix reordering approaches as

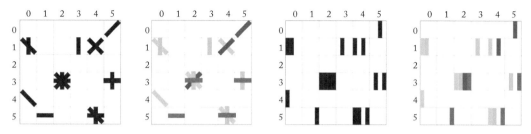

Figure 5.3: Four of the visual encodings used by Vogogias et al. [2020] for displaying multiple edge sets from different layers in a single matrix. Left to right: the encodings are orientation, color and orientation, position, and finally color and position. Following a user study and expert evaluation the authors' first recommendation is to use the combination of color and orientation.

they place rows sharing many interactions close to each other in the matrix. Such an approach to ordering may help to visually reveal communities of strongly connected nodes. Community detection in a bipartite graph consists in finding groups of nodes in one layer which are densely connected to groups of nodes found in the second layer (Task category **A** in Chapter 4). Two-way hierarchical clustering is commonly used with biological data for this purpose, e.g., Santamaría et al. [2008].

In presence of multiple layers, the comparison of link patterns between many pairs of layers may be useful to the analyst (Task category **A**). Laying out small multiples of matrix views side by side is one approach. Liu and Shen [2015] investigate several possible juxtaposition strategies, and assess their usability with multifaceted, time-varying networks. *MuxViz* [De Domenico et al., 2015] uses matrices to summarize layer-level statistics, as a means to convey a notion of layer similarity to the analyst (Task categories **A**, **B**, **D1**, and **D2**). In an evaluation by Alper et al. [2013], which focused specifically on weighted graphs in the context of brain connectivity analysis, matrix visualizations were shown to perform better for comparison tasks than node-link visualizations. Vogogias et al. [2020] present a method for encoding multiple edge types in a matrix (see Figure 5.3). The designs presented in this paper are aimed at Dynamic Bayesian Networks from a computational biology application but can be generalized to other multilayer networks. The designs use color, position, and orientation to encode on which layer the edge is located. The subsequent experiment found that orientation and color worked the best, but for local tasks orientation without color performed well.

There have not been a large number of techniques that have use matrices to visualize multi-layer networks. We elaborate further on this topic in challenges and opportunities (Section 5.7).

5.2   1D NODE-LINK REPRESENTATIONS

Many visualization techniques use one-dimensional representations of layers (see Figure 5.4 for examples of basic 1-dimensional visualization). This type of encoding can often leverage the

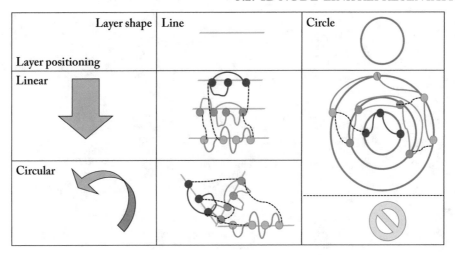

Figure 5.4: Illustrative examples of the types of 1D layout according to the layer shape (line vs. circle) and positioning of layers to each other (linear vs. circular). The layer shape restricts the positioning of inter-layer nodes. No examples were found in the literature of circular position combined with circular layer shape.

law of continuation of Gestalt theory, such that the eye may perceive paths on which nodes are arranged whether these paths are actually drawn or not. This applies to circular paths, as well as straight axes, or any curve shape.

5.2.1 LINEAR NODE-LINK REPRESENTATIONS

In this category a layer is represented by a straight 1D axis. The representation of a multilayer network lays out nodes on several parallel axes. An important way of distinguishing axis-based visualizations relates to the type of variable represented by the axis, whether it is quantitative, e.g., a graph metric like node degree or any numeric node attribute, or ordinal/ranking-based. Despite the visual similarity to parallel coordinates [Inselberg and Dimsdale, 1990], a polyline represents a path between nodes sitting in different layers/axes instead of a thread linking attribute values across different columns in a given table entry. Crnovrsanin et al. [2014] describe a view that uses such parallel axes arrangement and alternatively chord diagrams. Their example analyses consist of comparing "aggression networks" among students in four different schools. They show that smaller groups do not show internal aggression patterns, while larger groups victimize everybody equally (within the same group and in other groups). In this case the analyst is more interested in topological considerations at the group level, and structural differences between layers (Task category **D2**).

Ghani et al. [2013] provides an approach called *Parallel Node Link Bands*. Nodes are positioned uniformly across spaced parallel axes which represent layers defined by the node type

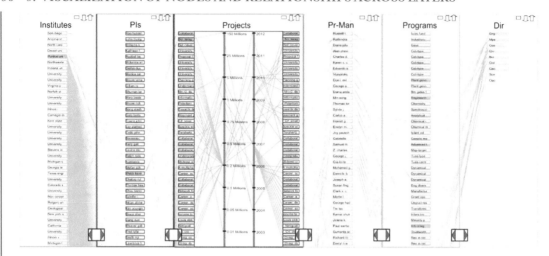

Figure 5.5: The *Parallel Node Link Bands* representation of Ghani et al. [2013]. Each axis is a distinct set of vertices. Edges are only displayed between adjacent axes. Some axes show a quantitative value, e.g., project budget, while others display text strings, sorted based on a graph metric or alphabetically.

(or mode); see Figure 5.5. Edges are only drawn between adjacent layers, and intra-layer edges are shown in a separate visualization. Edges present in non adjacent layers are not shown. Axes can be re-ordered to make any set pair of layers adjacent, showing the edges between them. Node order on axes can be set based on edge attributes or connectivity to other layers. Ghani et al. [2013] use their approach to analyze a data set concerning National Science Foundation (NSF) funding. Examples of tasks they carry out include determining whether some NSF program managers award funding to some PIs (Principal Investigators) more often than others on a three-layer networking containing program managers, projects, and PIs. This is an instance of Task category **A** where the focus is on paths traversing all layers.

The list view of the *Jigsaw* application of Stasko et al. [2008] provides an overview of entities grouped by type, with edges being drawn between connected entities in adjacent lists. One of the main utilities of this system is to relate different types of named entities (people, geographic locations, organizations) mentioned in the same documents. Entities which are connected to a currently selected item are highlighted by color across all lists. It therefore emphasizes the analysis of paths across all available layers (Task category **A**). The list view is complemented by a node-link and a matrix-like scatterplot view, among others.

One advantage of list based views is that they are an efficient use of space, and as such maybe a suitable choice for visualization where device constraints mean that display space is at a premium. The *Orchard* application of Eichmann et al. [2020] provides the ability to explore multivariate heterogeneous networks on mobile devices using a list-based paradigm. This ap-

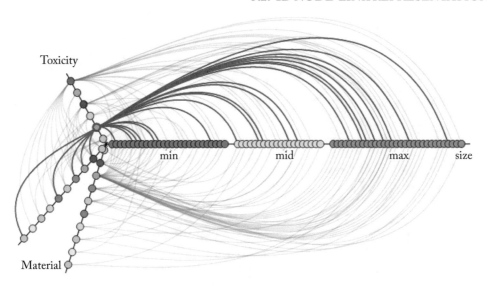

Figure 5.6: The hive plot representation of health data by Yang et al. [2016] showing four layers/axes: toxicity type, material (duplicated), and particle size. Edges are only displayed between adjacent axes. The vertices on the horizontal axis are colored based on their cluster membership.

proach facilitates the display of aggregated information concerning entities as bar charts in the list, and allows for gesture based interaction for exploration and navigation through the data set.

The Hive Plots of Krzywinski et al. [2011] differ from the previous techniques in that they arrange the axes radially. Originally introduced for the analysis of genomic data, they have been used in other domains such as performance tuning in distributed computing [Engle and Whalen, 2012], and in the domain of health [Yang et al., 2016], as can be seen in Figure 5.6. In Krzywinski et al. [2011], node (gene) subsets are placed on separate axes based on a node partitioning algorithm. The fundamental questions they answer using Hive Plots include determining differences in connectivity patterns between layers (Task category **D1**). An element's position along its axis is often calculated based on a graph metric, e.g., node degree in Engle and Whalen [2012] and may be based on the raw or normalized value of an attribute. Edges are displayed between adjacent axes only, however axes can be reordered to display the edges between pairs of desired axes. Yet, visual clutter may still occur with real application data. Axes may be duplicated to support the visualization of relationships between layers that are currently non-adjacent (Task category **C**).

5.2.2   CIRCULAR REPRESENTATIONS

Concentric circles are a type of circular representation, where each circle can represent a layer. Concentric circles are used by Bothorel et al. [2013] where the focus is on depicting paths

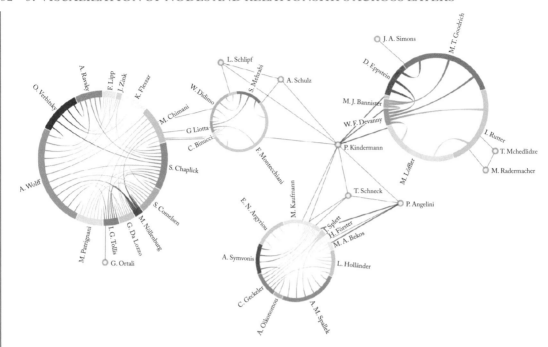

Figure 5.7: The chordlink approach of Angori et al. [2020] is a hybrid visualization approach that utilizes chord diagrams to replace clusters of nodes in a node-link diagram. The edges between nodes in a given set are drawn internally to the chord-link diagram. In a multilayer approach, layers could be represented by a chord diagram instead of clusters.

through the whole set of layers (Task category **A** in our taxonomy). Node order optimization and edge bundling [Holten, 2006] are used to reduce edge clutter. A similar layout is used in the ring view of the *MuxViz* application of De Domenico et al. [2015], but focuses on visual correlation analysis of node attributes across different layers (Task category **D1**). Node color encodes attribute values (see Chapter 7 and Figure 7.4), while ring order and ring thickness encode computed layer-level metrics. Similarly, *Circos* [Krzywinski et al., 2009] is a popular tool for comparative analysis of genomic data, where each ring/layer may stand for a biological sample. In order to compare node attribute values across samples, a histogram is wrapped around each ring (Task category **D1**).

Chord diagrams (see Figure 5.7 for an example) can display layers as arcs within one overall circle. They are used in the *NetworkAnalyst* tool of Xia et al. [2015] to analyze gene expression data. Links between layers are drawn as splines connecting identical nodes occurring in different layers/arcs (Task category **B**). The analyst may click on a pair of arcs to highlight their common nodes (and the bridging links). A similar approach is followed by Alsallakh et al. [2013] and Crnovrsanin et al. [2014]. In the presence of multilevel categorical attributes

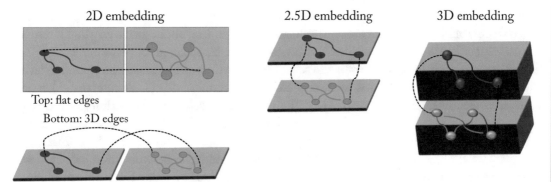

Figure 5.8: An illustrative example of 2D, 2.5D, and 3D visualizations of a network visualized using node-link diagrams with 2D and 3D layer shape. The 2D layout (left) is a standard node link layout where nodes are positioned with x and y coordinates and, usually, an orthogonal projection is used to render the visualization (top). In order to reduce edge crossing, sometimes a perspective rendering with 3D edges is used (bottom). The 2.5D approach (center) draws sets of nodes on 2D planes that are positioned in 3D space. Thus, the inter-layer edges are drawn in the third dimension. For the 3D example (right), nodes are at different depths in that they actually have an individual z coordinate (depth value) for position. In this example, the nodes themselves are 3D objects (spheres) rather than circles, and are shaded to clarify their 3D shape. 3D network visualizations are usually rendered with a 3D projection. As can be seen from this figure, depth is difficult to convey in a static 2D image. 3D visualizations usually require some level of camera movement to reveal occluded data, as well as the depth aspect to be conveyed through motion and/or stereoscopy.

as in the work of Humayoun et al. [2016], each arc of the chord diagram can further be split hierarchically (Task category **C**). The chords would then connect nodes at the leaf level across all layers where they are repeated.

5.3 2D LAYERS WITH NODE-LINK REPRESENTATIONS

Across the various papers we surveyed, node-link layouts cropped up frequently. As illustrated in Figure 5.8, for the standard 2D layout x and y coordinates are assigned to each node. For the 2.5D approach sets of nodes are assigned to planes at different depth, and for the 3D approach the nodes are usually assigned a different individual depth (z coordinate). The *MuxViz* toolkit of De Domenico et al. [2015], from the domain of complex systems, utilizes several variants of node-link visualizations. They are also used in other domains that depend on complex systems theory, as can be seen in the work of Gallotti and Barthelemy [2015], Bentley et al. [2016], and De Domenico [2017].

While not explicitly designed with multilayer network visualization in mind, constraint-based layouts offer the possibility to constrain a 2D node-link layout in such a way that it respects the concept of layers. For example, the *SetCola* constraint-based layout of Hoffswell et al. [2018] allows users to apply layout constraints to sets of nodes, which might easily correspond to layers. Such a layout approach supports analyzing cross layer connectivity (Task category **A**) as well as layer comparison (Task category **D2**). The examples covered by the authors include food web networks and a network modeling a biological cell, and both of these data sets can be considered to have multilayer characteristics. The multi-level layout approach of *TopoLayout* by Archambault et al. [2007] is also not designed with multilayer networks in mind. However, as an approach, it may be of interest in the multilayer case. *TopoLayout* is a feature based approach that decomposes a network into a hierarchy of features (hence, being considered a multi-level algorithm) and chooses a suitable algorithm to layout each feature. Within a multilayer network the structure of each layer may be very different, requiring an approach that can adapt to the different features in each. The highest level feature detected by the *TopoLayout* algorithm is a connected component, therefore the algorithm could be adapted to consider each layer as a connected component, but also would need to be enhanced further to consider the impact of inter-layer connectivity.

Inspired by the multi-level nature of some problem areas, e.g., biological networks, the 2.5D approach materializes layers as 2D parallel planes in a three dimensional layout, see Figure 5.8, or along a diagonal axis as done by Skrlj et al. [2019]. This visual design relies on the law of uniform connectedness of Gestalt theory. It separates intra-layer links lying within layer from inter-layer links providing a more natural support for path related tasks (Task categories **A** and **B**) than traditional 2D node-link layouts, but 3D navigation is required to allow the user to change their perspective on the data and resolve visual occlusion problems. As opposed to 1D axis-based representations, the parallel 2D planes provide space to lay out intra-layer links. In the 2.5D category, some approaches use color redundantly to encode layer information as in Fung et al. [2009] and Skrlj et al. [2019]. Another visual design option for 2.5D consists of using color to encode an attribute value or a computed metric, e.g., community assignment by a community detection algorithm as done by De Domenico et al. [2015], across the different layers.

5.4   JUXTAPOSITION VS. SUPERPOSITION OF 2D VISUALIZATIONS

For superposition, typically the color or shape of the nodes encodes the layer Fung et al. [2009], Kohlbacher et al. [2014], Moody et al. [2005], and Zeng and Battiston [2016]. Color can also be used for edges, e.g., De Domenico et al. [2015] and Ducruet [2017]. This design choice relies on the law of similarity of Gestalt theory (color similarity in this case). This design is often adopted when the multivariate nature of the network is the driving motivation of the visual design. For instance, Figure 5.9 represents flows of maritime traffic using color to encode different modes

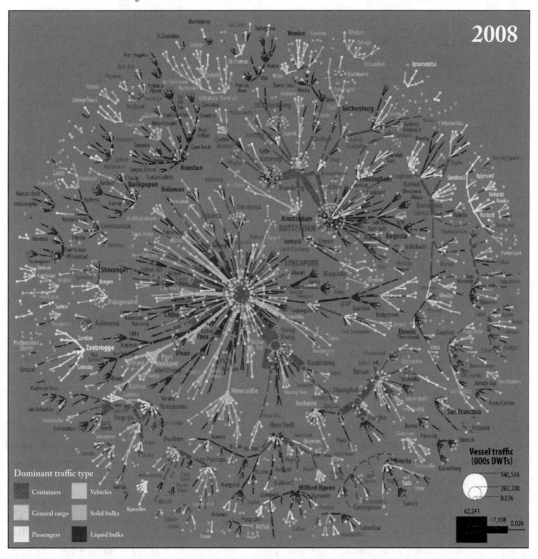

Figure 5.9: A multilayer network visualization describing the flow of maritime traffic. Nodes represent ports and different edge colors represent different modes of shipping, taken from Ducruet [2017].

of shipping (or layers). The analyst looks at, among other things, structural changes over time, where different layers encode different time slices (Task category **D2**). However, if the analyst is interested in analyzing a given time slice, different layers may represent different shipping modes. The related task consists of comparing structural differences among the different modes.

Juxtaposition   Superposition

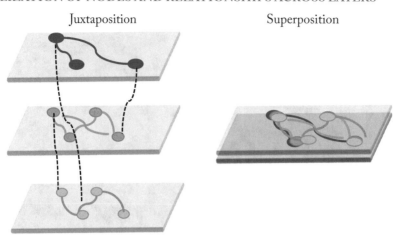

Figure 5.10: Two types of layer positions: juxtaposition and superposition.

In similar visual designs, layer information is diffuse, relationships between layers and within the same layer are mixed, and users seldom get a handle on layers to manipulate them directly. Nodes belonging to different layers are intertwined in the 2D plane, when standard node-link layouts are used, and edge clutter is problematic. Layer-related tasks may therefore be difficult to carry out with these representations.

A more widely accepted approach in information visualization, especially for the purpose of comparative analysis of graphs, consists of using small juxtaposed representations. This is often used for graph matching tasks, where the focus is on understanding commonalities and differences between a set of related networks, e.g., Hascoët and Dragicevic [2012]. In the context of this paper, the networks that need to be matched are distinct layers in a larger multilayer network (Task categories **D1** and **D2**). Whether in a 2.5D setting or in a flat small multiples setting, one challenge consists in ensuring that duplicate nodes are laid out consistently across layers, by introducing constrained layout strategies, as in Fung et al. [2009] and Hascoët and Dragicevic [2012], to better support cross layer entity comparison (Task category **B**). Node-link layouts have been used to compare networks visually (Andrews et al. [2009] and Di Giacomo et al. [2009]), but in many situations other approaches may be more suitable (see Sections 5.1 and 3.2.3).

More generally, juxtaposed views are often used in the domain of information visualization, and in many applications, e.g., the analysis of microarray data, as done by Santamaría et al. [2008]. In this case, 2D node-link views may be used as one of multiple complementary visualizations of a multilayer network, e.g., Ghani et al. [2013], Kairam et al. [2015], and Stasko et al. [2008]. It is yet possible to eschew the idea of using a node-link visualization altogether, for example using aggregate views of nodes (based on attribute data), such as a bar chart enhanced with arcs, as done in the *GraphTrail* application of Dunne et al. [2012]. Coordination between views is common, e.g., brushing and linking. The *Detangler* application of Renoust et al. [2015] builds on this by also harmonizing layouts between views. It supports several task categories identified in this survey, namely cross layer connectivity (Task category **A**), layer manipulation (Task category **C**), and layer comparison (Task categories **D1** and **D2**).

5.5   3D AND IMMERSIVE NODE-LINK REPRESENTATIONS

The use of 3D layouts is historically much less common in the information visualization research community. While some work has shown that there may be some benefit to 3D layouts, these results are only under stereoscopic viewing conditions; see Ware and Mitchell [2008]. Outside of stereoscopic viewing conditions, there are no empirical studies which demonstrate usability gains from a 3D graph visualization [Greffard et al., 2012]. However, at the time when much of this research was conducted, stereoscopic technology and head tracking were not as common or advanced as they are now.

Advances in display technologies mean that stereoscopic 3D visualization is becoming more relevant as a visualization approach, see Marriott et al. [2018b]. Augmented and virtual reality display hardware, such as the Microsoft *Hololens* and the *Occulus Rift* allow for fully stereoscopic head-tracked 3D visualizations. Kwon et al. [2016] used a fully immersive virtual reality headset to run a study on a 3D network visualization. In this study, graph nodes were rendered on the wall of a sphere surrounding the user, with the use of edge bundling [Holten, 2006] extended to route edges with depth. This approach was compared to a flat 2D visualization, also presented in the immersive environment. The results of their study suggested that there are benefits in terms of speed of user response. In addition, for larger graphs, users could give more significantly correct answers (a result consistent with Ware and Mitchell [2008]). Kotlarek et al. [2020] have also examined immersive 3D visualization, but focusing more on memory and graph structure. Their study used fully 3D layouts on a range of graph sizes and compared to a traditional 2D desktop environment. They found that immersive visualization of a network mitigated confusion caused by edge crossings and node overlap. These studies are not focused on multilayer network visualization, but they do indicate that immersive 3D visualization may be an approach to mitigate the difficulty in interpreting the complexity of multilayer networks in the future.

In the biological domain, the *Arena3D* application of Pavlopoulos et al. [2008] visualizes biological data using an interactive 3D layout, where layers are also projected onto planes, and entities are connected across layers by edges rendered as 3D tubes. The authors demonstrate its effectiveness by analyzing the relationship across layers, based on proteins and genes associated with a specific disease (Task category **A**).

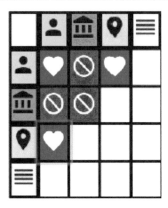

Figure 5.11: The preference matrix of Müller et al. [2017], which is used to specify which links should appear in the visualization, and also the preference level for edges to configure an underlying Degree of Interest function.

5.6   EDGE VISUALIZATION

The complex structure of multilayer graphs makes edge visualization an important consideration. Depending on the nature of the underlying graph, edges between layers may not be represented in the visualization. In Ducruet [2017] (see Figure 5.9), for example, edges between different layers are not represented since the relationship between nodes in different layers is that of identity, i.e., they are *coupling edges*, as the same nodes appear across the layers (multiplexity). In this case, showing all of the edges between different nodes in a single diagram provides a user information about the relative amounts of different modes of traffic between ports. *Detangler* [Renoust et al., 2015] only displays all intra-layer edges in a single graph, an edge between nodes of the secondary layer graph indicates that both layers bear at least a same intra-layer edge. In other cases, there may be different types of links between layers and it may be important in some cases to distinguish between inter-layer and intra-layer links. Or, it is possible that the number of layers may cause enough clutter with respect to edges, and the resulting visualization becomes less understandable.

In some cases, the chosen solution is to simply not draw all edges and to allow the user to choose which edges to see via interaction to ease inter-layer comparisons (Task category **B**). For example, the *Parallel Node Link Bands* technique of Ghani et al. [2013] only draws inter-layer edges between nodes on parallel axes, and intra-layer edges are displayed in a separate visualization. In the work of Müller et al. [2017], interaction is also used to select which type of relationships are to be displayed in the visualization. A user clicks on a preference matrix to specify which links should appear in the visualization, allowing multiple edge types (i.e., inter-layer edges) to be included in the displayed visualization without drawing them all and overwhelming the user.

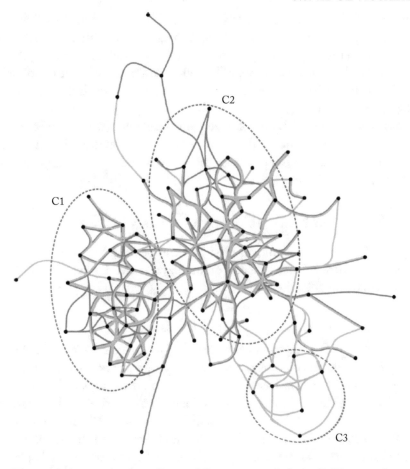

Figure 5.12: The multilayer edge bundling of Bourqui et al. [2016]. The graph is the *Reality Mining* data set where edges are interactions between mobile phones. Layers are encoded using color. The three clusters (C1, C2, and C3) represent clusters of users in the data set.

The technique of edge bundling [Holten, 2006] is well established and has been shown empirically to be beneficial for understanding higher-level network structure; see McGee and Dingliana [2012]. Bourqui et al. [2016] have adapted it for the multilayer use case. The authors bundle all edges in a single visualization, in an aesthetically pleasing manner, with edges being kept adjacent to each other when they share a common path, and edge crossing being avoided (see Figure 5.12). This approach is useful for showing edges from multiple layers in a single visualization (where there is no division of nodes between layers); the approach is agnostic to the source or target layer, or whether the edges are inter- or intra-layer edges (Task categories **A** and **B**).

Within their list-based view, Crnovrsanin et al. [2014] use edge bundling between different list columns as a clutter reduction technique to clarify similarities between different edge types. The authors essentially group edges based on relation type, by clustering the vertices and altering the clustering based on vertex mode. They also use a modified edge bundling in their circular layout, that distinguishes intra-layer edges and inter-layer (which they refer to as between-mode) edges; see Figure 5.13.

Quite naturally, the visualization of edge group structures (see Vehlow et al. [2017] for a survey) presents a lot of similarities with multilayer network visualization and some work cited can be easily adapted especially for cross layer connectivity and layer reconfiguration (Task categories **A** and **C**).

5.7   CHALLENGES AND OPPORTUNITIES

A wide range of visualization techniques exist that can be used for, or adapted to, visualizing multilayer networks. There are many aspects of multilayer network visualization that are opportunities for immediate investigation.

5.7.1   MATRIX VISUALIZATION APPROACHES

Matrices have a number of advantages over node link diagrams, particularly when the main tasks of interest only involve direct adjacency; the principal advantage being a near elimination of visual clutter. Evaluations of matrix-based approaches have existed for some time (e.g., Ghoniem et al. [2005b]). Despite this and more recent evaluations of matrix-based approaches for visualization (e.g., Okoe et al. [2018] and the work of Nobre et al. [2020] focusing on multivariate networks), the visualization literature generally has under-explored the use of matrices for applications visualizing networks and multilayer networks are no exception. The only evaluation of the use of of matrices for multilayer network visualization is the work of Vogogias et al. [2020]. We would encourage future researchers to explore the use of matrices for multilayer network visualization, and to further evaluate their benefits for multilayer network visualization tasks (discussed in Chapter 4) .

5.7.2   HYBRID APPROACHES

Recent work has explored the integration of multiple visualization techniques in an effort to better grasp underlying data [Javed and Elmqvist, 2012]. Although matrices have been shown superior to node-link diagrams for dense networks, the latter may better facilitate the tracing of paths through multiple nodes. In this spirit, *NodeTrix* from Henry et al. [2007], mixes node-link views with matrix-based visualizations to support typically locally dense social networks. While *NodeTrix* is not explicitly a multilayer network visualization technique, it is the first hybrid approach that focused specifically on network visualization. Since its inception, the idea has been extended by other techniques to support other types of data, such as compound graphs;

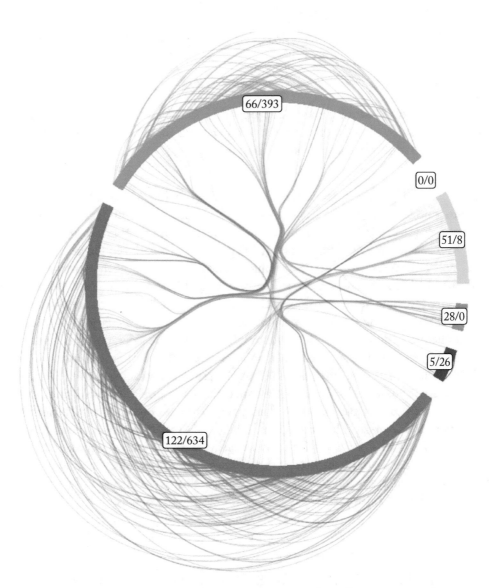

Figure 5.13: Edge bundling as utilized by Crnovrsanin et al. [2014]. Intra-layer edges are routed around the exterior of the circle. Inter-layer edges are routed via the interior of the circle and bundled.

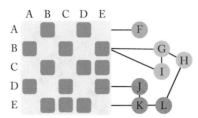

Figure 5.14: A simple illustrative example of a hybrid approach, mixing both matrix and node-link in a single visualization. In this case the purple layer is more dense in terms of intra-layer links, and may benefit from being visualized as a matrix.

see Rufiange et al. [2012]. Incidentally, it is worth noting that hybrid visualizations of matrices incorporating node-link concepts has been shown to improve path-finding tasks as evaluated by Sansen et al. [2015]. Recent work has also considered chord diagrams instead of adjacency matrices in this style of hybrid visualization (Angori et al. [2020]; see Figure 5.7). Although they generally have not focused on visualizing multilayer networks, such approaches could be adapted to do so. Different layers may be visualized using different techniques depending on the structure of the layer and its relationships with other layers. A hybrid approach may even be used within a layer visualization. Different visualization techniques can be considered depending on the local structure within a layer. They even might take into consideration the edges between nodes in a given layer and other layers (be they inter-layer edges to different nodes).

With respect to using a different visualization approach for each layer, *VisLink* by Collins and Carpendale [2007], allows visualizing a data set using multiple representations at once. It also explicitly displays the cross-view links. Using the technique, one layer could be used for each representation, and inter-layer links could be highlighted (Task category **A**). Adopting another perspective, *HybridVis* by Liu et al. [2017], allows using the same kind of representation, but for different levels of details (or hierarchical scales). In this case, a node-link view may include some levels that are shown as expanded, and other levels are shown as collapsed (Task category **C**). With additional views (histograms, parallel coordinates), more details on layer level attributes can also be obtained (Task categories **D1** and **D2**).

5.7.3   IMMERSIVE VISUALIZATION

As discussed in Section 5.5, immersive visualization may also offer opportunities for multilayer network visualization given recent advances in display technology (see Marriott et al. [2018a]). The 3D may offer reduced occlusion when compared to two dimensions. This may allow for simultaneous visualization of multiple layers and their connectivity, allowing for an overview of a complexity structure without losing the detail of the layer contents. The work of Kotlarek et al. [2020] and Kwon et al. [2016] focused on immersive visualizations which use Virtual Reality. Augmented Reality (AR) devices are also increasing in popularity and capability. They frequently are part of situated visualization systems, which are visualizations that are related to, and displayed in, their source environment, see Thomas et al. [2018] and White and Feiner [2009]. If the popularity of AR technology continues to increase, use of immersive 3D for the visualization of multilayer networks may be desirable to allow users to inspect data at a specific site related to the network, alongside AR views of the site. A key issue with visualization of data in an immersive environment is that end users find interacting with the data much easier in a traditional desktop environment. As shown in the study of Bach et al. [2018], participants subjectively reported more difficulty in perceiving and interacting with data in an immersive AR environment compared to a traditional desktop, with the authors explicitly noting that interaction in the immersive environment was more cumbersome. This is understandable due to the many years of research that have supported familiarity with visualizations based on the desktop metaphor. Consumer-level immersive technologies are a recent development, and as Bach et al. [2018] note, training can lead to further performance improvement in immersive AR. Currently, there are no immersive visualizations that focus explicitly on multilayer networks. However there are techniques for 3D graph visualization that could be applied to the multilayer case. For example, the use of implicit surfaces to define a bounding area for a set of nodes (as done by Balzer and Deussen [2007]), may prove useful for defining the boundaries of layers in a 3D immersive multilayer network visualization.

5.7.4   EDGE ROUTING

The need to address tasks related to cross layer entity comparison also means that there may be interesting opportunities with respect to edge routing and visualization. The approach used by Crnovrsanin et al. [2014] is not developed much beyond the original edge bundling algorithm of Holten [2006], while the bundling of Bourqui et al. [2016] focuses on edge routing in the case where the nodes and edges of all layers are presented in a single node-link diagram. Edge bundling has already been shown to be useful for comparison of hierarchical structures, by Holten and Van Wijk [2008]. Adapting it futher for the multilayer use case may give insights about relationship structure across a set of layers. The force-directed edge bundling approach of Holten and Van Wijk [2009] provides more flexibility than the original approach of Holten [2006], as it does not require a node hierarchy, however the resulting bundling is dependent on node position. Adapting such an algorithm for a multilayer network may require different considerations and constraints to be applied to inter-layer and intra-layer edges. Edge Bundling has become a popular topic within information visualization, and the survey of Lhuillier et al. [2017] describes a range of approaches and applications which may be of interest to those who are looking to apply edge bundling to multilayer networks.

5.7.5   ENCODING OF ENTITIES

Using other approaches to the encoding of entities other than position and color may be an avenue for further research. Enclosure is a principle of Gestalt theory and a visual cue described by Mackinlay [1986]. It is used by approaches such as *bubble sets* by Collins et al. [2009] and *Gmap* by Gansner et al. [2010] to indicate set or cluster membership.

Use of the concept of enclosure can be seen in the *BicOverlapper* system of Santamaría et al. [2008], which uses biclustering methods to find relationships between groups of genes and related groups of medical conditions. It uses coordinated multiple views, one of which employs convex hulls, similar to the approach of Collins et al. [2009], within a standard node-link representation, to represent groups of genes. These overlapping convex hulls are meant to support the identification of commonalities and differences between layers (Task categories **B** and **D2**). Generally, many set visualization techniques (see Alsallakh et al. [2016] for a survey) may be relevant to the visualization of multilayer networks.

5.7.6   OTHER COMPLEX NETWORK VISUALIZATION APPROACHES

Within this book we have intentionally avoided focusing on more complex data modeling approaches such as hyper-graphs. However, it is worth noting that in many applications, particularly within biology, data sets are explicitly modeled as hyper-graphs, e.g., the *Systems Biology Graph Notation* of Le Novere et al. [2009] that is often used to describe biological pathways. Representing hyper-edges in a multilayer context (particularly if endpoints belong to discrete layers) is an interesting open challenge.

5.7.7   DYNAMIC AND TEMPORAL LAYERING

Some multilayer data sets also contain a temporal aspect, e.g., Gallotti and Barthelemy [2015], and there has been much work done in the field of complex systems on the dynamics of multilayer networks, see Boccaletti et al. [2014]. However, integration between temporal and other aspects for dynamic multilayer networks may still offer opportunities for novel visualization techniques.

CHAPTER 6

Interacting with and Analyzing Multilayer Networks

Contents

Since 1996, analyzing and interacting with networks has been driven by the well-known Visual Information-Seeking mantra of Shneiderman [1996] (*Overview First, Zoom and Filter, then Details on Demand*). However, this mantra is purely data oriented without the use of any model on the data. The continuous growth in the size and complexity of the data makes it difficult to use nowadays, especially the "Overview first" step which often leads to the so-called hairball effect. To support this expansion in size and complexity, Keim et al. [2008] proposed the Visual Analytics mantra (*Analyze First—Show the Important, Zoom, Filter and Analyze further, Details on Demand*). It introduces the notion of a model on top of the data to be able to first show and interact with what is important. However, both mantras follow the same principle, which is to analyze and interact from a global scale to a detailed scale, and eventually find interesting details.

This approach (global scale to detailed scale) is not usable in every domain (e.g., in social sciences or life sciences). Sometimes, data experts have incomplete knowledge of their data because it is too voluminous, too complex or too recent, or a global representation of the data is simply not possible. Experts may also be unable to simply conceptualize what is an overview of their data. It is then necessary to be able to explore the data without a perfectly defined initial goal, in particular to validate the potential of the data to answer the questions formulated by the experts. Usually, experts have focused and specific questions about their field, which the handling of data (detailed scale) allows, or even requires, to be put into context (global scale) with specific interactions. This approach can be seen as a new mantra as published by Luciani et al. [2019]: *Details First, Show Context, Overview Last*. To our knowledge, it has hardly been investigated by the visualization community so far except with a Degree of Interest approach, see Van Ham and Perer [2009], and multivariate network exploration allowing users to exploit both network structure and data, see Van den Elzen and Van Wijk [2014].

Since multilayer networks bring more structure with the notions of layers and aspects, we believe multilayer networks are a good candidate model to support this new mantra effectively. Layers are able to bring the seeking context to the analysis as in Laumond et al. [2019]. Following the *Science of Interaction* from Pike et al. [2009], layers make a good reasoning aid for end-users. We introduce layer-based interactions in Section 6.1, which are either low-level (on network structure) or high-level (with the data). Interactions are traditionally used as the visual interface for analytic measures, and we discuss such measures for multilayer networks in Section 6.2. We conclude this chapter by describing challenges and opportunities related to interacting with and analyzing multilayer networks (Section 6.3).

6.1   INTERACTION APPROACHES

Based on observation of user behavior, Amar et al. [2005] have detailed a set of low-level analytic tasks in support of user reasoning, centered on numerical operation and data selection: *Retrieve Value, Filter, Compute Derived Value, Find Extremum, Sort, Determine Range, Characterize Distribution, Find Anomalies, Cluster,* and *Correlate*. Although not all these tasks operate on the same level (arguably, *Cluster* could be a combination of *Compute Derived Value, Retrieve,* and *Filter*). These low-level tasks can be considered in the context of user manipulations of the data from a visual perspective, i.e., interaction. Yi et al. [2007] grounded a first categorization of interactions organized around user intent, proposing seven categories: *Select, Explore, Reconfigure, Encode, Abstract/Elaborate, Filter,* and *Connect*.

These categories, whether low or high level, are agnostic to the data structure, and therefore also concern interactions of networks, and multilayer networks. Interaction may be supported at different levels, from individual network elements (e.g., individual nodes and links) to the level of whole layers—whether single layers or groups of layers (Task category **C1** in the task taxonomy of Chapter 4.1). This is well illustrated by Hascoët and Dragicevic [2012] in their system, called *Donatien*, which supports the *reconfigure* and *explore* interactions of Yi et al. [2007].

6.1.1   LAYER-LEVEL INTERACTIONS

Interacting with entire layers is one unique feature of multilayer networks and is also discussed as part of the layer-level tasks described earlier in Section 4.1 (Task categories **D1** and **D2**). Layer-level interaction helps structuring the context of the multilayer network. It does so by operating at its highest level of abstraction. Depending on the visual analytics approach taken,

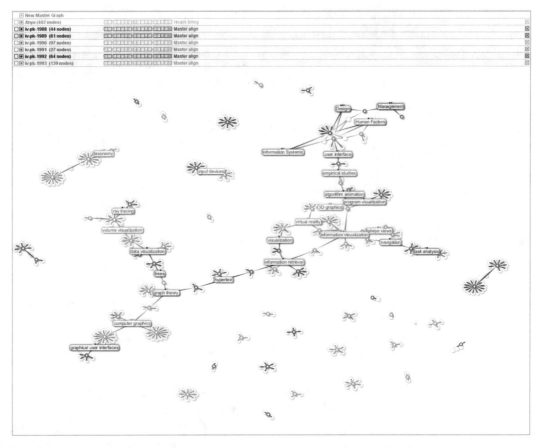

Figure 6.1: A screen shot from *Donatien* [Hascoët and Dragicevic, 2012] showing layer stacking with color coding.

it supports *Overview first* in the mantra of Shneiderman [1996] should it come first, or *Show Context*, and *Overview Last* from the more recent approach of Luciani et al. [2019].

The *Donatien* system of Hascoët and Dragicevic [2012] offers three different spatial organizations of layers: (1) small multiples, (2) stacking the layers on top of each other, and (3) animation. Starting from the small multiples view, the analyst can drag and drop a layer onto another one, to stack them and more easily compare their elements based on the distinctive layer color. In the stacked mode, a set of title bars provides an affordance to reorder the layers in the stack interactively (Figure 6.1). The title bars also include reconfiguration tools, e.g., choices of layout algorithms that are either applied to the layer being manipulated or to the whole stack of layers. The set of title bars can be interacted with to achieve flip-book animation, which can be useful for comparison across layers. This seems quite a natural approach when the layers are

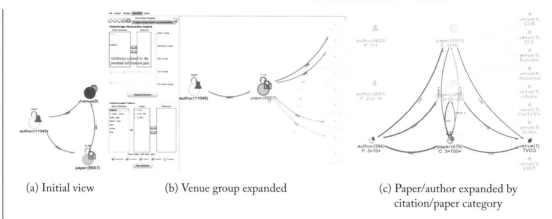

(a) Initial view   (b) Venue group expanded   (c) Paper/author expanded by
                                                                citation/paper category

Figure 6.2: Focus+context in *OnionGraph* [Shi et al., 2020].

defined as consecutive snapshots of a dynamic network. Also in the stacked mode, *Donatien* clusters nodes from different layers together based on their spatial proximity in the pixel space. The analyst is able to edit the resulting clusters interactively by pulling a node out, or by dragging and dropping a node onto another node (or group of nodes) to merge them. Merged nodes carry a color coded pictogram relating them to the layers they occur in.

More structured layer organizations, such as a hierarchy of layers, may prove to be necessary. This ensues from the concepts of aspects and layers put forth by Kivelä et al. [2014], but also from many real application needs. From an interaction perspective, merging layers together or splitting them apart (Task Category **C1**) becomes a matter of manipulating their parent node in that layer hierarchy. In relation to this concept, the *Ontovis* system of Shen et al. [2006] uses an ontology visualization to steer the associated network visualization. An ontology could be seen as an artifact representing the layer structure of a multilayer network. In a simpler manner, *MultiNets*, from Piškorec et al. [2015], provides different views of multilayer networks such as geographical and force layout views with overlaid edges, and interactions to select the layers and node/edge properties to display.

The *OnionGraph* application, from Shi et al. [2020], provides a hierarchical focus and context approach targeted specifically toward heterogeneous data (Figure 6.2). The hierarchy provides different levels of abstraction based on node type, role equivalence, and structural equivalence. In their example use case using an academic publication data set, the heterogeneity of the data is derived from node types, and edges only exist between certain node types. There is no formal layer definition, and the abstraction used to provide the hierarchical focus and context is applied across all data types, and does not fully consider the heterogeneity of the data. Such abstractions could be adapted to be applied on a per layer basis. This could be very useful in multilayer systems, particularly for comparison of complex layers.

The *Hub2Go* interaction approach [Zimmer et al., 2017] supports heterogeneous network exploration by automatic camera movements in multiple network views to facilitate the navigation across interconnected networks. Colored links between single networks show relationships between them. They are enhanced by numerical values for the connection type, such as how many authors in a publication network are connected with selected universities in an affiliation network.

The *Orchard* application of Eichmann et al. [2020] is also an application for exploring multivariate heterogeneous networks, however it focuses on mobile devices, requiring a touch based approach to interaction. The application employs a list-based visualization approach. Users use touch to interact and a swipe gesture to navigate between different node types (referred to as pivoting by the authors).

6.1.2   NODE-LEVEL AND LINK-LEVEL INTERACTIONS

Traditional modes of interactions include:

- *selection*: point and click selection, lasso selection of nodes;

- *filtering*: keeping/removing nodes or links based on attribute values; and

- *navigation*: visually inspecting a fragment of the visual representation using zoom and pan, or context+detail techniques (e.g., fisheye distortion, magic lenses, or detailed node/edge inspection with hovering and tooltips).

Interaction at the level of nodes and links typically operate at the finest detailed representation of the multilayer network (the *Details on Demand* of Shneiderman [1996] or *Details First* of Luciani et al. [2019]). While this family of interaction helps direct investigations of all elements of a graph, it also supports users in forming a mental-map of the complex problem observed, by iteratively making sense of its larger context (Luciani et al. [2019]'s *Show Context*).

These have obviously been used widely with standard node-link representations, and are directly applicable one layer at a time in the context of multilayer networks. A few approaches have however enriched traditional node- and link-level interactions.

Hascoët and Dragicevic [2012] argue that layer level interactions require a visual affordance [Norman, 2013]. In the *Detangler* approach of Renoust et al. [2015], layers are abstracted into a different graph. It is coordinated to offer this new affordance from selection of layers and groups of layers, with coordination to the node- and link-level interaction. *Detangler* combines two distinct synchronized visual representations. A first panel (Figure 6.3, left) displays the overall network connectivity through a node-link view between nodes of all layers. Another panel (Figure 6.3, right) displays a different node-link view which abstracts how layers interact. In this view, nodes correspond to layers of the multilayer network and links are determined whenever layers overlap. *Detangler* offers two asymmetric coordinations (corresponding to Task category **A**). A first selection from nodes and links (left panel) highlights the corresponding

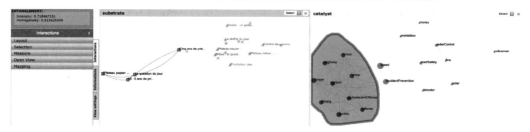

Figure 6.3: A screen shot from *Detangler* [Renoust et al., 2015] showing how nodes (left panel) relate to layers (right panel). Selecting layers (lasso) triggers the highlight of nodes they involve (red nodes, left panel).

layers they belong to (right panel). Another selection from the layers (right panel) highlights all nodes and links shared between all layers, or within at least a layer (determined by a Boolean operator). This asymmetric coordination can be further expanded through "leapfrogging" (executed by double-clicking the selection lasso), piping the two coordinations together. This *expands* the original selection to include all nodes (Boolean OR) involved in any one of the layers that were selected; or *restricts* the original selection to nodes involved in all layers that were selected from the layer view (Boolean AND). A specific layout further supports these interactions because shared nodes and links are located close in the same layout area of the layers they belong to.

As shown by Hascoët and Dragicevic [2012], small multiples make an excellent ground for comparison tasks in multilayer networks (Task category **B**). Although linked-highlighted small multiples can instantly reveal common entities between two layers of a multilayer graph, when spatial layouts are not aligned between the layers it becomes difficult to identify individual or groups of common elements. Renoust et al. [2019b] handle this limitation by offering a *drag-and-drop* interaction between two views (Figure 6.4). From a user selection in a source view, they provide an animated interpolation upon the drag interaction in the target view, which enables users to follow the desired group of elements. The animation can be further controlled (rewind or fast-forward) by the location of the mouse in the target view.

6.2   MULTILAYER ANALYTICS

When it comes to the analysis of multilayer networks, common metrics, community detection algorithms, and centralities that are applied to standard networks may be applied to a single layer in isolation, not considering its relationships with other layers. However, this is ignoring the impact of the interplay and relationships between layers, and the higher level patterns that layer coupling can offer, which provide one of the key motivating factors for adopting a multilayer approach in the first place. Fortunately, many of the analytic techniques have been adapted for the multilayer case, and new techniques specific to multilayer networks have emerged.

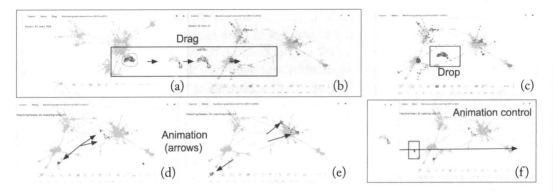

Figure 6.4: Animated drag and drop (a,b,c) from Renoust et al. [2019b]. The animation (d,e) projects the dragged elements onto the target layer, and the animation can be controlled with the mouse (f).

De Domenico et al. [2013] provide a mathematical formulation of multilayer networks within a tensorial framework and discuss the generalization of many standard metrics, centralities, and a generalized community detection for use in multilayer network analysis. There are also many analytic techniques to be found applied to the types of network that fall within the general multilayer framework. Battiston et al. [2014] describe measures for the analysis of multiplex networks. They not only generalize standard measures but also consider global properties, for instance the shortest paths between nodes and across multiple types of layers. They also consider the multiplexity, which was first introduced by Podolny and Baron [1997] and corresponds to the number of coupled edges in a graph.

The wide range of existing literature on k-partite networks also provides techniques which may be useful for the multilayer case. When it comes to comparing nodes between two different layers, or the relationship between a pair of layers, bipartite analysis techniques come in handy. Borgatti and Everett [1997] provide bipartite versions of measures that could be adapted to analyze the relationships of nodes between two layers. An example of a bipartite specific measure which may be adapted to the multilayer case is the *redundancy coefficient* of Latapy et al. [2008]. It measures the fraction of pairs of neighbors of a node in a bipartite graph which are connected by a different node. It could be considered to show the importance of a node in connecting two layers in a multilayer context.

Some pre-existing network-related challenges can also benefit from a multilayer analytics approach. For example, Liben-Nowell and Kleinberg [2007] describe that an analyst may try to determine whether or not are there going to be new interactions/connections between two people in a near future. This is known as link prediction, i.e., the prediction of previously unobserved or unknown links in a network. The link prediction problem has been explored in an explicitly multilayer context by Pamfil et al. [2020], who examine the inference of edge correlation across multiple layers to improve the quality of predictions.

There exists a great number of techniques that can be used for the analysis of multilayer graphs from the complex systems community and other domains, such as social network analysis. In their seminal work on the definition of a framework for multilayer networks, Kivelä et al. [2014] provide a review of many techniques which have been generalized from standard single layer approaches to the multilayer case. However, within information visualization the adaptation of these techniques is still rather limited. The most prominent example is the work of Renoust et al. [2015], described earlier in this chapter, which utilizes multilayer analytics in support of cross layer selection (mainly, edge entanglement later explored in Škrlj and Renoust [2020]). More recently, Melançon et al. [2020] support dynamic multilayer analytics with the interactive coordination of node-link diagrams, Sankey diagrams, and traditional time plots based on multilayer network metrics for subset definitions, plots, and community detection, with tracking over time. Ghani et al. [2013] consider bipartite versions of centralities in their work on multi-modal network visualization, but this simply reflects that their analysis often only considers two modes of data. The *MuxViz* tool of De Domenico et al. [2015] makes extensive use of multilayer analytics. This is not surprising as it is a multilayer analytics visualization tool from the domain of complex networks. However, its aim is explicitly to showcase these techniques and visualize the results, rather than use the techniques to help an end user within the context of a larger visual analytics application. *Py3plex* of Skrlj et al. [2019] combines a diagonal layout for visualization of multilayer networks (emphasizing the edges between layers), accompanied with network aggregation, similarity, and multiple measure computation functionalities. In the same line, other tools have combined some form of basic multilayer visualization and analytics—one can mention *Pymnet* from Kivelä et al. [2014], *Multinet* from Dickison et al. [2016a], or also *multiNetX* from Amato et al. [2017]. These all make analysis and visualization available from their API, but no interaction is specifically designed to manipulate layers or data, or designed with higher level analysis in mind.

6.3   CHALLENGES AND OPPORTUNITIES

Multilayer network related tasks and exploration may require novel interaction techniques. *Detangler* is one example of an interaction technique to support multilayer network exploration (Task categories **A**, **C**, and **D**). The *Donatien* application of Hascoët and Dragicevic [2012] supports interaction techniques related to comparison of multiple layers (Task category **D2**), and defining layers for comparison (Task category **C**). However, there still is a large design space to be explored for multilayer use cases, particularly exploration between layers (Task category **D**) and layer creation/manipulation (Task categories **C1** and **C2**).

6.3.1   INTERACTIVE QUERYING AND LAYER DEFINITION

While some application domains inherently provide layer definitions (for example genomics, metabolomics, proteomics, etc. are useful default layer definitions in systems biology), defining and refining layers that are useful to the end user is an important task that requires user interaction with the data. Indeed, without an initial layering, a network cannot actually be considered as a multilayered network. This comes slightly before our tasks taxonomy (see Chapter 4) which considers networks with already defined layers. However, as mentioned in Section 4.3, we could consider an initial network without layers as a multilayer network with only a single layer. Interaction design following Task categories **C1** and **C2** would support this initial layering. Creating such a layering should consider the potential complexity and the size of the multilayer network. For instance, some data sets may require a visual query interface to allow a user to find a data subset of interest that may become a candidate layer.

The notion of querying a database by example has existed for some time (as far back as Zloof [1977]), and more recently visual query based approaches allow users to specify a visual example and query graph data sources without knowing the specific data entities they are looking for. *Visage*, by Pienta et al. [2016], is an early example of such an approach, allowing users create queries by sub-networks from node types or specified nodes. More recently, *Graphiti*, from Srinivasan et al. [2018], allows users to interactively specify example subsets of nodes and links to create unipartite networks from a multivariate data set. When joining multiple data sets, "graph wrangling" is a useful approach that allows users to define nodes, edges, and their classes. *Origraph*, by Bigelow et al. [2019], is one very good example of graph wrangling, realized for traditional graphs. The use of a meta overview makes it very intuitive and could be naturally extended to layer and aspect definition.

Visual query approaches have more recently been extended to multilayer data sets. Cuenca et al. [2018] describe a tool that searches a large multilayer graph for specific patterns of data (specified visually by a node-link drawing). It displays the matching patterns using a heat-map of the overall network, as well as Kelp-like diagrams (inspired by Dinkla et al. [2012]) that convey the patterns in the original graph. A more recent version of the tool named *VERTIGo* [Cuenca et al., 2021] provides additional functionality. It provides a visual list of query results, in the form of node-link diagrams of the returned sub-graphs, referred to as *embeddings* by the authors. It also suggests new edges to be added to a query pattern, integrating the suggestions as part of the original visual query. A user can add new edges that link nodes that were part of the original query, referred to as *internal edges*, or edges to nodes that were not included in the original query, referred to as *external edges*. The internal edges are added by clicking on dashed links between nodes in the query suggestion. The external edges are added by clicking on segments of a pie chart glyph the represents a node in the query. Each segment represents a different external edge type and the size of the segment represents the proportion of the already discovered embeddings that can be expanded with the associated edge type. The layer definition used by *VERTIGo* is based purely on multiple relationship types, i.e., multiplexity. It does not consider layers defined by any other means, or allow for any user definition of layers.

McGee et al. [2019b] also provide a visual interface to query large multilayer data sets, providing an interactive query builder using dynamic radial menus that reflect the structure of the underlying data set. While this approaches helps users to find a subset of interest in a large multilayer data set, the layer specification (based on node and edge types) is already defined within the data. Interactive visual querying for layer definition remains an open challenge.

6.3.2   MERGING, SPLITTING, PROJECTING, AND TRANSFORMING LAYERS AS INTERACTIONS

One way that new layers can be defined is by interactions with the existing ones, combining them or splitting them into smaller layers (Task category **C1**) or deriving new layers based on projection and transformations (Task category **C2**). These manipulations are helpful to modify a multilayer network such that it would better fit a user's scope of analysis. For example, consider a bipartite author-paper network, where researchers are connected to papers that they authored. A projection on the paper node-type results in a co-authorship network of researchers, where two researchers are connected if they ever authored a paper together. Bipartite networks can be transformed by projecting on a node type, and analyzed using bipartite metrics such as re-dundancy [Latapy et al., 2008]. Such an operation may be applied to a multilayer use case, as discussed by Interdonato et al. [2020].

Simplification falls under this category and Interdonato et al. [2020] provided a taxonomy of techniques (including projection) to simplify a multilayer network while making layers more relevant to the users' goals. To design such specific interactions that enable an easy and free manipulations of layers, it would be useful to consider a visual approach using layers as entities (a meta view such as the network model in *Origraph*, from Bigelow et al. [2019], might be a good lead into it). Visually manipulating aspects could also prove useful to this challenge. Some interaction techniques from multivariate network visualization, see Van den Elzen and Van Wijk [2014] may also be useful in refining definitions of aspects and their associated layers.

6.3.3   INTERACTIVE LAYER COMPARISON

When visually analyzing a multilayer network with a large number of layers and/or aspects, a user may wish to compare layers or their associated attributes. The order in which layers are presented may impact the users ability to understand the similarities and differences across layers. The in-teractive sorting of layers is an open challenge. Within multivariate visualization approaches, such as parallel coordinate plots, the sorting of axes impacts the users ability to recognize pat-terns. The impact of sorting can be seen in the multimodal network visualization of Ghani et al. [2013], where relationships are only seen between adjacent axes of the *Parallel Node Link Bands*, and interacting with the axis order (or duplicating axes) is a key element of using Hive Plots (see Chapter 5). Ordering may be controlled from a layer network, as introduced in *Detangler*

[Renoust et al., 2015] as well. If a higher-dimensional visualization approach (2D, 2.5D, or 3D) is taken to visualize layers and their constituent entities, interactively sorting and ordering layers still faces many of the same challenges, and further issues related to clutter and occlusion.

As described previously, the *Detangler* application of Renoust et al. [2015] uses node color and positioning to support cross-layer selection, as well as linked highlighting to relate entities across two layers. It is clear that the visual encoding of entities selected by interactions across layers is important, and as mentioned in Section 5.7, the encoding of entities is an open research opportunity. We advise further examination of the Gestalt principles in action to show the impact of cross layer interaction techniques. For example the principle of "common fate," where entities are perceived as a set if they move together, may support interactions by which multiple layers receive a new layout.

6.3.4   EXPLOITATION OF EXISTING MULTILAYER ANALYSIS TECHNIQUES

As stated in our discussion of multilayer analytics in Section 6.2, there is a wide range of analytics tools and techniques which could be used in the visual analysis of multilayer networks. The lack of application of these techniques within existing visual analytics tools—a challenge that has been identified early on by Rossi and Magnani [2015]—can be seen as an opportunity for the research and development of new tools and techniques that can provide end users with the answer to their questions.

CHAPTER 7

Attribute Visualization and Multilayer Networks

Contents

Network data may have *attributes associated with nodes and/or edges*. The possibly multiple attributes may be categorical or numerical. For example, in social networks attributes such as age, gender, or location may describe person nodes, and attributes such as duration of social contact may be attached to edges describing relationships between persons. Specific characteristics of nodes or edges may be used to create layers, e.g., the contact type (via telephone or email), or an individual person may create layers by describing contacts across different types. The same may apply to node characteristics because a person's occupation may create layers of contacts between employees or between managers. The resulting multilayered networks are usually still multivariate, as persons and the represented relationships between them carry additional attributes that describe both of them. These attributes need to be analyzed together with the layered network structure. For example, it may be of interest to analyze the age and gender of persons having contacts at employee level and at manager level. Is there a difference between age groups at different network layers? Is there a difference in network structure between contacts of younger vs. older persons? Are some contacts via telephone taking longer than personal meetings, and if yes which ones?

Moreover, these networks may change over time, i.e., they are *dynamic*. The time dimension creates an additional aspect for a multilayer network—additional layers created by the time moments. Across the time aspect, both the network structure and the node and edge attributes may change. For example, the duration of telephone calls between persons may increase over time.

As for single-layer networks, each layer of the multilayer network can be described by standard network measures (i.e., statistics or metrics), such as density or average node centrality. These form attributes that are associated to the whole network at each layer. Such *network-associated metrics* may be treated as attributes to be analyzed and compared across layers. For example, people may prefer to telephone conversation meetings in person, and thus the telephone network layer is less dense than the in-person layer.

Tasks associated with multivariate attributes of multilayer networks may be answered with suitable visualization of multivariate, multilayered networks. However, the design of multivariate network visualization is a challenge even when a network consists of a single layer, as discussed in Kerren et al. [2014a]. The multiple layers pose additional difficulties to the visualization and interaction design. In this chapter, we examine existing techniques for visualizing multivariate data across multiple layers. We structure the chapter according to the type of attributes and conclude with an overview of existing challenges and research opportunities.

7.1   NUMERIC ATTRIBUTES

Simple approaches involve the complete separation of the attribute visualization from the graph structure. To better relate the relationship between networks structure and attributes, the attributes may be integrated into the network visualization itself (referred to as augmented network visualization by Dickison et al. [2016a] and Rossi and Magnani [2015]) or a linked view brushing approach maybe taken (i.e.,, by using multiple coordinated views), by which the relevant related nodes would be highlighted in a network view, when selected in the attribute visualization and vice-versa (Task category **B** in the task taxonomy of Chapter 4). Separating or integrating attribute views from/into network views needs not be a binary decision. The *MVN-Reduce* method of Martins et al. [2017] combines these two perspectives into a single unified model. This model includes both attribute- and edge-based similarities, with a user-controlled trade-off. Consequently, the network nodes are projected into a 2D embedding with the help of a dimensionality reduction algorithm and their neighborhood varies depending on whether the user is more interested in the network topology or the attributes. Extending this idea to multiple layers, e.g., by adding similarities between network elements in different and/or more distant layers, would be surely beneficial.

The standard multivariate visualization by parallel coordinates is also a suitable basic visualization technique. In the case where the graph is multiplex, and nodes appear in all layers, the different axes can represent a specific layer attribute. Heatmaps may also be adapted for a multilayer use case. For example, the temporal heat maps by Grottel et al. [2014] could be made suitable for multilayer attribute visualization, by using graph layers instead of time slices for each column (see Figure 7.1).

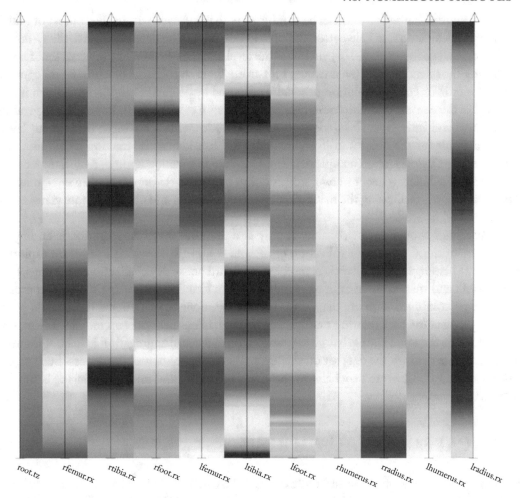

root.tz   rfemur.rx   rtibia.rx   rfoot.rx   lfemur.rx   ltibia.rx   lfoot.rx   rhumerus.rx   rradius.rx   lhumerus.rx   lradius.rx

Figure 7.1: The temporal heat map of Grottel et al. [2014] showing changes in attribute values over time slices in the vertical direction.

The approach used for attribute visualization relies heavily on the task the user is performing. For example, a scatterplot matrix is one technique by which attributes may be summarized, possibly even across layers. However, if the user's goal is to understand correlations of attributes across layers, an approach such as a modified multilayer version of the *scatterplot staircase* (SP-LOS) proposed by Viau et al. [2010] may be more efficient in terms of comprehension and space. In this approach, scatterplots of the attributes are ordered pairwise based on correlation and common axes.

Attribute visualization also can be combined with *interaction* within the context of multilayer graph visualization, to help better understand the connection between layers. The *Detangler* application of Renoust et al. [2015] visualizes the level of entanglement of a selected set of nodes by coloring the selection lasso (an attribute measuring internal cohesion of a group—as opposed to group inertia or entropy, a notion originally proposed by Shannon [1948], and which is also proposed in the context of multiplex networks by Battiston et al. [2014]). In general, navigating across intra- and inter-layer links/edges to adjacent nodes within the same or different layers may be difficult for the user, even if no attributes are considered. Interaction techniques may help to reduce this navigation complexity. The Hub2Go interaction approach of Zimmer et al. [2017] supports heterogeneous network exploration by automatic camera movements in multiple network views, facilitating the navigation from and to nodes and other attribute views across interconnected networks.

Attributes should not be considered only at a per node level. Aggregation is an important feature of the *GraphTrail* application of Dunne et al. [2012], which focuses on exploring multivariate heterogeneous networks. It eschews standard network visualization encodings, such as node-link and matrix, in favor of aggregate attribute visualizations using a hybrid approach of bar charts combined with arc diagrams. The list-based views of *Orchard* [Eichmann et al., 2020], an application for exploring multivariate heterogeneous networks on mobile devices, also aggregate node data into simple bar visualizations. Such a visual aggregation is helpful for users as they navigate through the data set on the limited screen space of mobile devices. Such aggregation based approaches are beneficial to the characterization and understanding of layers and their interactions.

7.2 CATEGORICAL ATTRIBUTES

An interesting example of categorical multivariate data, which could be adapted for multilayer visualization can be seen in the multivariate graph analysis tool of Pretorius and Van Wijk [2008].

Their approach uses icicle plots to describe the (hierarchical) categorical attributes of the source and target of a set of directed edges. The source icicle plot is on the left side of the screen and the one for the target nodes is on the right, with the edge and their associated data drawn in the middle; see Figure 7.2. Such an approach may be easily adapted to compare categorical labels across layers (Task category **B**). Combined with edge bundling, as done by Holten and Van Wijk [2008] using *Hierarchical Edge Bundles* Holten [2006], it could also be used to examine structural and categorical attribute difference between layers simultaneously (Task categories **B** and **D2**).

It is also possible to consider categorical attribute data as a network layer in and of itself. For example, in the application *OntoVis*, Shen et al. [2006] use a node ontology graph to query a large heterogeneous social network data set. The node ontology graph reflects the disparity of the attribute (how well distributed it is across nodes), and its edges display frequency of links

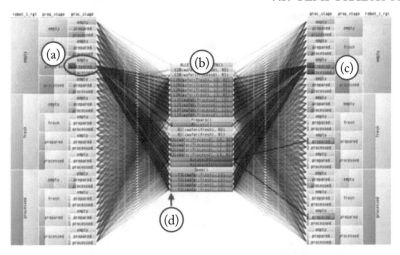

Figure 7.2: A screenshot taken from the multivariate network visualization work of Pretorius and Van Wijk [2008]. The icicle plots on either side of the screen represent the hierarchically structured categorical data of nodes, and are connected by multivariate edges. The annotations (a), (b), (c), and (d) are related to user tasks in the original work.

between the entities. It acts both as a visualization of aggregated categorical data, and a layer by which the data set can be better interacted with and understood.

7.3   TEMPORAL ATTRIBUTES

Multivariate multilayer networks, within which one aspect defining layers is time, can be considered a form of multivariate dynamic graphs. This visualization is challenging as both changes in network structure and in network attributes over time need to be shown possibly across multiple additional layers. The visualization of dynamic networks with multiple attributes often employs node-link diagrams. The visualization of the temporal aspect often relies on animation or small multiples as described by Archambault et al. [2014] and Beck et al. [2017]. Each time step of the network together with the associated attributes can be shown using one of the above-mentioned visualization techniques. However, these approaches may lead to cognitive overload with difficulties to keep the so-called mental map (i.e., the stability of the drawing) during the exploration of the data. Therefore, specialized techniques have been developed that aim at showing multiple layers, time, and the attributes in one view, without the need to compare multiple views or remember animated view sequences. This, however, leads to more complex visualizations. One example is *SmallMultiPiles* by Bach et al. [2015], where the temporal aspect is shown as an overlay in a matrix-based visualization. This is able to show edge weights. Several layers can be shown as several matrix rows next to each other. If the network layers are structured in a hierarchical way and edges have associated weights, the *TimeRadarTrees* of Burch and Diehl [2008]

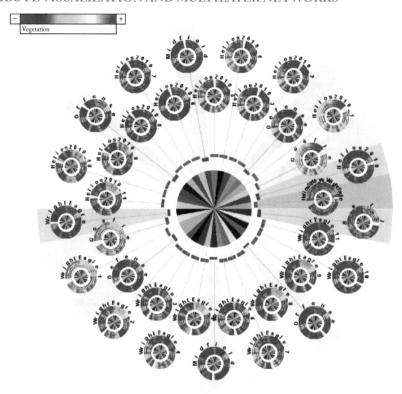

Figure 7.3: A screenshot taken from the approach Layered TimeRadarTrees by Burch et al. [2011].

can be used for showing dynamics across layers and edge weights. This is a nested ring-formed visualization with edge weights shown as color in a ring-formed heatmap and the network structure shown in circular layouts outside the main ring visualization. This technique is a basis for a multilayer version called *Layered TimeRadarTrees*, by Burch et al. [2011]. The layers are placed along several rings; see Figure 7.3. A less cluttered version of the visualization with fewer edge crossings makes use of circular layouts—*TimeSpiderTrees* by Burch et al. [2010]. A different approach is to separate the visualization of network structure over time and network attributes over time in linked views, as done by Ma et al. [2015]. Some approaches show attributes within matrix-based visualization and then use animation to show the time development (for instance, see Berger et al. [2019], Rufiange and Melançon [2014], and Yi et al. [2010]). This approach can be extended for multilayered dynamic graphs with attributes, when the so-called superadjacency matrix, as referred to by Kivelä et al. [2014], is used for showing the edge structure across layers. An interactive way of dealing with scalability of dynamic networks is to use Degree-of-Interest functions to filter out only nodes and edges of interest to the user; see Abello et al. [2013]. This can also include temporal filtering.

7.4   NETWORK-ASSOCIATED METRICS AS ATTRIBUTES

Attributes need to be considered across layers, and attribute values (especially for numerical attributes) may change across layers, especially if the attributes are derived from graph metrics, which may be calculated on a per layer basis. An example of this can be seen in the *MuxViz* toolkit of De Domenico et al. [2015]. Here, the authors use an annular ring visualization approach, which shows the values of metrics across layers, with each ring representing a layer, or in some cases a different centrality for a specific layer; see Figure 7.4 (Task category **D1** or **D2**). As described in Section 3.2.3, attribute-based visualization can also be used to summarize the structure of a network and allow for network (or layer) comparison, as done by *ManyNets* from Freire et al. [2010].

7.5   CHALLENGES AND OPPORTUNITIES

Attribute visualization in multilayer networks is important, for instance, for understanding the differences of a node's attribute values in different layers or seeing attribute patterns considering the network topology at the same layer (see Task categories **D1** and **D2**).

7.5.1   MIXED ATTRIBUTES ACROSS MULTIPLE LAYERS

In the previous section, we discussed three typical data types and properties that usually occur in practice. The presented example visualization approaches, however, focused mainly on single attribute types. In real-world data sets, this is often not the case, i.e., attributes types are mixed together and must be analyzed together. Let us think about multilayer social networks. Here, numerical attributes might be the ages of people, their salaries, and any kind of numerical rating; their categorical attributes might consist of their addresses, names, titles, etc.; and a temporal attribute could be the start time of their activities (even if the network under investigation is not changing its topology, the attached attributes might change over time). The interactive visualization of this situation of mixed attributes is highly challenging and still an important open research question in general, but especially so for multilayer networks.

7.5.2   MULTILAYER NETWORK CENTRALITIES AS ATTRIBUTES

When we restrict ourselves again on numerical attributes, many existing techniques can be adapted to the multilayer case, as seen above with tools such as *Detangler* by Renoust et al. [2015]. Many classical network centralities have been adapted for multilayer networks; see Domenico et al. [2013] and Kivelä et al. [2014]. While *MuxViz* by De Domenico et al. [2015] does include the visualization of centralities, there is much opportunity for novel visu-

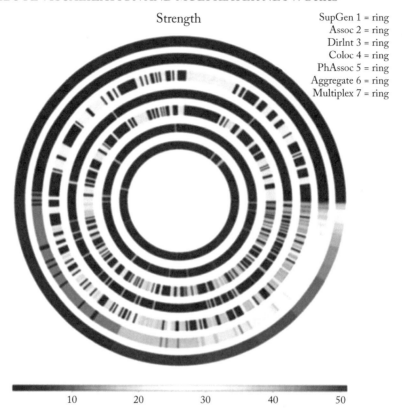

Figure 7.4: **A** screenshot taken from MuxViz [De Domenico et al., 2015] shows the values for a centrality measure across seven layers. Each ring stands for a layer, numbered from the center outwards. Nodes repeat across all layers, where angle indicates node identity. Each node is materialized by a ring segment whose color corresponds to the centrality value of the node. Nodes can be sorted according to their attribute value in a reference layer, for instance layer 7 (the outermost layer) in this case. Node ordering helps reveal correlations across layers. For example, rings 3 and 5 are negatively correlated, while rings 6 and 7 are positively correlated. It should be noted that this visualization approach affords more space to the attributes of nodes encoded in the outer rings, an issue that would be avoided with a linear representation.

alizations considering specific multilayer centralities, integrated into network visualizations for supporting cross-layer comparison based on both attributes and network topology.

7.5.3   NOVEL ALTERNATIVE REPRESENTATIONS

In the previous section, we already introduced the visual designs of *separating* and/or *integrating* attribute views from/into the actual network views. According to Jusufi [2013] and Kerren et al. [2014b], there are even more possibilities to solve the problem of visualizing such multivariate networks: *semantic substrates*, *attribute-driven layouts*, and *hybrid approaches*. Semantic substrates are simply "non-overlapping regions in which node placement is based on node attributes," an approach first proposed by Shneiderman and Aris [2006], who combined it with sliders to control the edge visibility and hence the visual clutter. Consequently, the underlying graph topology is not (always and/or completely) visible. Attribute-driven layouts arrange nodes and edges in the visual display to give insight about the attached multivariate data instead of visualizing the underlying graph topology, i.e., the attributes become now the first-class citizens in the final visual representation. In contrast to semantic substrates, this approach does not inevitably use non-overlapping regions, but controls the node placement in the graph layout by using its attributes. Hybrid approaches combine at least two of the previously discussed techniques. The most common combinations are multiple coordinated views with any of the integrated approaches. For instance, *JauntyNets* by Jusufi et al. [2013] collates coordinated views with an attribute-driven layout. Attribute values and a two-dimensional embedding (produced with a so-called dimensionality reduction method) of the multidimensional attributes are shown in individual coordinated views. The main view is based on an attribute-driven network layout. The attributes (or groups of them) are arranged in the shape of boxes on a circle whereas the network itself is positioned in the inside of this circle. A force-based system is established between the network nodes (as is standard in force-based graph layouts) but also between the nodes and the attribute boxes. Depending on the strength of the relationship between a node and an individual attribute, the nodes are placed closer or more distant to the attribute box; see Figure 7.5.

All of these approaches were designed for multivariate networks, but most of them can be enhanced or modified for the visual analysis of multilayer networks, for example, by introducing additional forces and representative visual objects for layers in case of the *JauntyNets* approach.

7.5.4   MULTILAYER NETWORK EMBEDDINGS

Finally, we want to highlight opportunities in the area of *network embeddings* (see Grover and Leskovec [2016] or Goyal and Ferrara [2018]) and their visualization. A network embedding projects network nodes to a low-dimensional representation while preserving the network structure; this is very similar to dimensionality reduction for multidimensional data and may incorporate supervised or unsupervised machine learning approaches. There is an increasing trend toward this relatively new network analysis paradigm, as described and surveyed by Cui et al. [2019]. There are embedding methods for both multivariate and heterogeneous networks, i.e., they consider attributes, network topology, and various node/edge types (sometimes referred to as "network embeddings with side information" [Cui et al., 2019]) in their embeddings. However, techniques that cover many layers explicitly by introducing within-layer and cross-layer node similarities also based on existing attributes, such as the *AMLNE* framework proposed by Pei et al. [2018], are still rare and not much in the focus of visualization research yet.

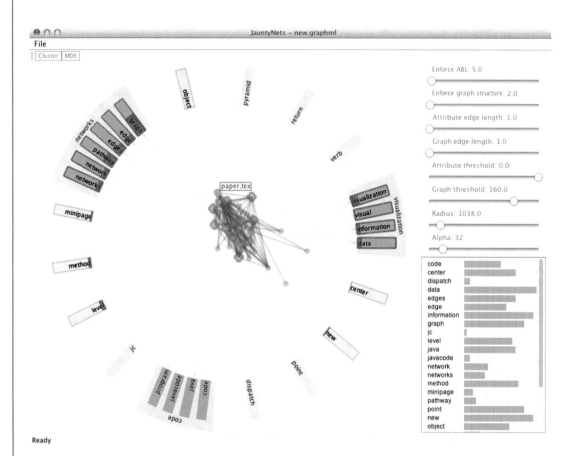

Figure 7.5: The *JauntyNets* tool. The graph view on the left shows three attribute groups on the outer circle and the network as a node-link diagram inside of it (based on a publication data set). Attribute nodes act as bar charts for the highlighted node in green, giving insight into the values for each of its attributes. Taken from Jusufi et al. [2013].

CHAPTER 8

Evaluation of Multilayer Network Visualization Systems and Techniques

Contents

From the user's point of view, the understanding of the visual display of multilayer networks is a challenging task. Multilayer networks are very complex resulting in a significant amount of cognitive load on the users. It is an open question how to visually design the representations of such complex networks so that users can derive useful insights from such visualizations. Our goal in this chapter on evaluation is twofold. First, we want to show that results from evaluation studies and cognitive psychology can be applied or re-used to improve the visual design of representations of multilayer networks. For this, we highlight existing evaluation examples and relevant psychological research. Note that our aim in this work is not to provide a comprehensive overview of this topic, but to discuss a few study examples to indicate the validity of our position. Second, based on such previous research, we derive some first tentative recommendations and also identify open issues that need to be addressed in future research to clarify how multilayer networks should be visually represented. Our goals are also relevant from a methodological point of view, because a theoretical framework based on previous research allows for a more rigorous design of experiments. Moreover, a theoretical framework enables researchers to conduct experiments more systematically and to generalize their results beyond a single evaluation study. This chapter is based on the paper by Pohl and Kerren [2019].

The potentially large design space of suitable visual representation and interaction techniques (see Chapter 5) has a direct influence on how visualizations of multilayer networks are perceived by humans. In consequence, different visualizations may lead to discrepancies with respect to cognition. Figure 8.1 conceptualizes three potential ways to visualize a multilayer network by using stacking in 2.5D/3D, 2D nesting of layers, and 2D alignment (e.g., by juxta-

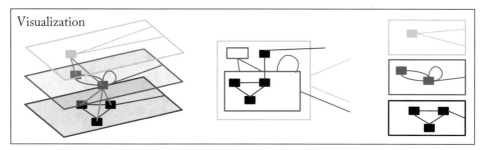

Figure 8.1: The examples of multilayer network visualizations. Different shades of color separate different layers. Taken and adapted from Schreiber et al. [2014].

position). Of specific importance is the visual representation of the intra- and inter-layer links of a multilayer network (see also Chapters 3 and 5). *Intra-layer* links show the relationships (links) between nodes in the network within the same layer and *inter-layer links* the relationships between the nodes of networks within different layers. If the network becomes larger, the differentiation of the edge types causes additional clutter beyond the usual clutter that we have when visualizing large networks in general; see Kerren and Schreiber [2014]. In addition, navigating across intra- and inter-layer links to adjacent nodes within the same or different layers may be difficult for the user. Interaction techniques may help to reduce this navigation complexity. Examples are *Bring&Go* by Moscovich et al. [2009], which moves adjacent nodes into the current view, close to the actual selected node, or *Hub2Go* by Zimmer et al. [2017], which supports heterogeneous network exploration by automatic camera movements in multiple network views to facilitate the navigation from and to nodes across interconnected networks. For concrete visualization techniques and approaches to displaying networks, such as layouts of node-link diagrams or matrix-based approaches, we refer the reader to Chapter 5 and to the graph drawing and network visualization literature: Di Battista et al. [1999], Herman et al. [2000], Kaufmann and Wagner [2003], Kerren and Schreiber [2014], Kerren et al. [2014a], Tamassia [2013], Vehlow et al. [2017], and von Landesberger et al. [2011].

Many of the multilayer network visualization papers from the information visualization domain described in this book are either system papers or design studies. Their evaluation frequently involves user feedback, such as done in Dunne et al. [2012] and Ghani et al. [2013], visualization expert reviews, as done by Shi et al. [2014], or usage scenarios as described by Dunne et al. [2012] and Renoust et al. [2015]. Currently, there is a lack of low-level empirical evaluations specific to multilayer network visualizations. This is because only a few low-level tasks have been defined in this context (see Chapter 4). Another problem is the absence of existing visualization and/or interaction techniques to directly compare with. Therefore, an empirical comparison of the user performance based on specific low-level tasks is often not practical. Within research fields beyond information visualization and visual analytics, there is less demand (or no tradition) for performing a thorough evaluation of novel systems or techniques. In such cases, the authors may just demonstrate the techniques with a suitable data set, e.g., Crnovrsanin et al. [2014].

8.1   OVERVIEW OF EXISTING EVALUATION STUDIES

As there are still very few studies doing serious evaluations of multilayer networks as described in the previous paragraph, it might be useful to look at studies of other complex networks (e.g., Burch et al. [2021] and Yoghourdjian et al. [2018]) and try to transfer the results of these studies to the domain of multilayer networks to better inform the design of multilayer network visualizations.

Yoghourdjian et al. [2018] provide a comprehensive overview of research on large and complex networks with a focus on evaluation studies. They point out that it is still an open question how to define large or complex graphs, but that most visualization studies were done with smaller graphs, and only very few with graphs with more than 1,000 nodes. While graph size has been recognized as an important factor for quite some time, per von Landesberger et al. [2011], there is no widely accepted definition of what is a large or small graph, with many empirical evaluations running pilot studies or basing their chosen size of large graph on assumptions rather than a standardized definition. Yoghourdjian et al. [2018] mention that other measures than number of nodes could also be used (number of edges, density), but these numbers are reported less often in the literature. They also found that there is a relationship between tasks, interactions and application areas, and basic measures of size of node-link diagrams. Large node-link diagrams are mainly used for overview tasks, as opposed to smaller node-link diagrams. Interaction techniques appropriate for larger, more complex node-link diagrams are different than the ones for smaller node-link diagrams in that in larger node-link diagrams interaction techniques requiring less effort are predominantly used. The authors argue that more detailed knowledge about cognitive scalability would be beneficial for the design of usable large and complex node-link diagrams.

Yoghourdjian et al. [2018] also point out that there are different possibilities to simplify sensemaking processes with large graphs. One possibility is to use *aggregation*, and the other is to offer appropriate *interaction* possibilities to the users. While interaction is an important approach to ease the cognitive load of the users, Hegarty [2011] points out that interaction also has a cost because it puts the burden of choosing the specific appearance of the visualization (choice of variables shown, choice of segment of the visualization shown, etc.) on the users. This requires some degree of meta-knowledge that not all users possess. In the following paragraphs, we briefly discuss contributions addressing the issue of aggregation and interaction, and we also look at evaluations that are specifically targeted at matrix representations.

Aggregation   Rossi and Magnani [2015] discuss design considerations for multiplex networks of social actors. They assume that special designs for the visualization of multiplex networks are necessary. While their paper does not present an evaluation, it addresses important design issues. The authors argue that one possibility to achieve an improvement of multiplex network visualization is by adding analytical measures (e.g., degree distribution). They discuss several different possibilities to ease the cognitive load on the user, e.g., by slicing the node-link diagram into layers with similar edges and showing them as small multiples (either with the same or with different layout). The solution they find most promising is to show only edges conforming to a relevance measure. In this way, they try to identify hidden clusters in social networks.

Simplification of large and complex node-link diagrams has been discussed to some extent in the scientific literature. Dunne and Shneiderman [2013] propose to achieve this simplification through abstraction with glyphs. They argue that in some application areas (e.g., social networks) certain motifs are meaningful ways to represent the underlying structure of networks. They suggest the use of fans (nodes with a single neighbor), connectors (linking a set of nodes) and cliques (completely connected nodes) as motifs. In a user study, they show that these abstractions are helpful for many types of tasks. They also provide guidelines for the design of the glyphs. Nevertheless, the authors point out that these motifs have to be learned and therefore increase the cognitive load. In addition, it is not yet clear whether there are other glyphs which might be more appropriate for other domains.

Archambault et al. [2010] investigate path-preserving clustering of graphs resulting in opaque meta-nodes. Their results are not entirely unambiguous. Nevertheless, they state that these results imply that path-preserving clusters can lead to an improved performance on global tasks.

Yoghourdjian et al. [2018] introduce the notion of thumbnails: small icon-like visualizations representing high-level structures of graphs (these are discussed in the context of layer comparison and overview in Chapter 3). Thumbnails are presented as small multiples to allow users to compare these high-level structures. The authors point out that there has been some research to investigate detailed comparison of individual changes, while their research addresses comparison of more general characteristics of graphs. They investigated in detail the possible design alternatives for thumbnails and came up with circular structures that also allow the inclusion of annotations of the visualization. In general, thumbnails were more effective than either node-link diagrams or matrices for overview comparison of large graphs.

Interaction   Another possibility to support the users' sensemaking processes is to offer interaction features (see also Chapter 6). Ware and Bobrow [2005] use highlighting to enable users to get meaningful insights from node-link diagrams of varying sizes (from 32 nodes up to 3,200 nodes). The results indicate that highlighting is a fairly efficient interaction method, especially for large networks. The authors state that without highlighting error rates were rather high even for the smallest network. They also show that search times were rather high for the largest network even with highlighting. This is an indication that large graphs are a form of visualization posing specific challenges.

Nekrasovski et al. [2006] compared pan and zoom interfaces with a rubber sheet navigation (as an example for focus and context methods). They found out that, contrary to their expectation, pan and zoom interfaces were significantly faster than rubber sheet navigation. Added overviews did not contribute to the success of the participants, however, the participants subjectively appreciated the overviews nevertheless.

Marner et al. [2014] developed an interaction technique for graph representations on large screens. They used animation to support users in the processes of rearranging nodes on the screen. They found that animation can be helpful to support users in this activity. They also point out that the results of evaluation studies for small graphs cannot always be transferred easily to the design of large graphs.

Other investigations discussed the importance of clustering and annotation (Abello et al. [2006], Archambault et al. [2011], and Archambault et al. [2008]) for large graphs. The goal of these interaction possibilities is to reveal meaningful patterns in an overwhelming amount of data.

Matrix representations   While node-link diagrams are the most prevalent form of network visualization in terms of use and evaluation, as discussed in Chapter 5, matrix representations can also be beneficial for network visualization. The work of Ghoniem et al. [2005b] showed that matrix visualizations were superior for some simple network tasks, with the notable exception of path tracing. It is worth mentioning that later work by Sansen et al. [2015] proposes and evaluates hybrid solutions to improve performance at this task when using matrix based visualizations. Okoe et al. [2018] showed that matrix visualizations were superior for some cluster related tasks. Most recently, Nobre et al. [2020] examined matrix visualization in the context of multivariate network visualization. Their work validated many existing claims and showed that matrices supported cluster based tasks with attributes, tasks that benefit from sorted attributes, and tasks that require scanning an entire network in the presence of attributes than are not relevant to the task at hand. They also concluded that matrix representation were at least as good as node-link representations for tasks related to edge attributes. None of the preceding matrix representation evaluations focused explicitly on multilayer networks. The only such evaluation so far is that of Vogogias et al. [2020] which focused explicitly on encoding multiple edge sets (such as the edge sets in different layers) in a single matrix representation (see Chapter 5 for more information). Further evaluations of matrix representations in the context of multilayer networks are required, particularly with respect to the tasks discussed in Chapter 4.

8.2   OVERVIEW OF RELEVANT PSYCHOLOGICAL RESEARCH

There are several areas in psychology that might be relevant for evaluation studies of multilayer networks, especially in the area of cognitive load; see Huang et al. [2009]. In general, it is necessary to take cognitive issues into account and use cognitive models as a foundation for the design of visualizations; see Hegarty [2011].

Cognitive load   As mentioned above, increased cognitive load is probably the most serious challenge facing designers of multilayer network interfaces. It has been argued that restricting investigations of visualizations to simple tasks can be misleading. Cognitive load theory can form a theoretical foundation for getting a more comprehensive picture of the cognitive processes users engage in when interacting with complex visualizations.

Cognitive load theory has been developed to describe the cognitive processes that learners engage in when interacting with educational material. Sweller [1988] distinguishes between *intrinsic* and *extrinsic* cognitive load. Intrinsic cognitive load results from the nature of the material presented to learners; extrinsic cognitive load results from the manner in which the material is presented. Both types add up to the total cognitive load. If this exceeds the working memory resources, the learners will not be able to process the information presented to them. Sweller et al. [2011] have also described various effects based on cognitive load theory that can be used as a framework for an improved design of educational material.

One example of such an effect that might be relevant for information visualization is the so-called *split-attention effect*, [Sweller et al., 2011]. This effect occurs when users have to attend to at least two pieces of information that are separated either in time or in space. To make sense of this information, the user has to integrate them in a meaningful way which requires a considerable cognitive effort, because some pieces of information have to be kept in short-term memory. Sweller et al. [2011] suggest some ways to alleviate this problem, some of which are also relevant for the design of information visualizations. They especially point out that elements that interact with each other should be presented in an integrated way. When designing visualizations of complex and large multilayer networks, it seems to be obvious to separate the single layers so that users are not overwhelmed by the information. Nevertheless, in that case users lose the information about the relationships of nodes to other layers. Designers should be aware that these relationships have to be indicated clearly, so that the cognitive load to remember these relationships does not become too heavy.

Cognitive load theory has been developed for educational purposes but is also applicable in other domains. It has been applied to the design of information visualizations. Huang et al. [2009] highlight that an inappropriate design of a visualization may lead to an increased extrinsic cognitive load. Changing the form of the visualization can alleviate this problem. High cognitive load occurs when many elements of the visualization have to be processed simultaneously. Huang et al. [2009] developed a cognitive load model for the evaluation of information visualizations and tested it successfully. They especially show that mental effort is an important aspect in interacting with visualizations. To a certain extent, users are able to counteract increased complexity of visualizations by exerting mental effort. In their experiment in the context of network visualizations, they showed that even relatively small node-link diagrams (25 nodes and 98 edges) can impose a high cognitive load on users. The authors point out that it is still not entirely clear which factors induce cognitive load. Most experiments in evaluation of visualizations have been conducted with fairly simple tasks. In realistic situations, complex tasks requiring an increased mental effort play a more important role.

Cognitive limits   Cognitive load theory implies that there are limits in cognitive resources. Franconeri [2013] discusses 15 of such limits in our ability to process visual information. He distinguishes between limits on identification of objects and limits on object selection. When these limits are exceeded, response time will increase and accuracy will decrease. Franconeri studies several cases in which such phenomena will occur. In some cases, it is for example difficult to find objects among a number of distractor objects. Another example for limits in object identification is inattentional blindness (users engaging in a demanding task will miss important information right in front of their eyes). Limits on object selection also affect the sense for location of objects. From five up to eight locations can be selected at the same time. Detecting the relative spatial relationships of an object is also a very demanding task. Knowledge about such limitations have to be taken into account when designing visualizations.

8.3   RECOMMENDATIONS FOR THE DESIGN OF MULTILAYER NETWORKS VISUALIZATIONS

Based on the literature reviewed above, some tentative recommendations can be provided for the design of multilayer networks. Large and complex graph visualizations in general and multilayer networks in particular have not been evaluated extensively so far. Therefore, these recommendations can only be a starting point. They also indicate areas where future research is necessary as listed in the following.

Our first three recommendations address HCI-related issues. Yoghourdjian et al. [2018] argue that tasks and interaction techniques for larger and smaller node-link diagrams differ. The authors also point out that aggregation and interaction are important techniques for supporting sensemaking processes with large node-link diagrams.

R1. *Task evaluation:* There is some indication that large node-link diagrams support different tasks and interaction possibilities than small ones. Nevertheless, this assumption is not entirely clear, and detailed evaluations in this area have to be conducted.

R2. *Aggregation:* Different forms of aggregation are possible. Aggregation in general is helpful, especially for overview tasks and detection of the general structure. The results so far are promising, but not yet conclusive. All the suggested forms of aggregation have been developed for specific application areas, and it is an open question whether these results can be generalized.

R3. *Interaction:* Highlighting, pan and zoom, and animation have been found to be helpful for supporting sensemaking processes with large networks. It is an open question whether other forms of interaction are also useful.

Recommendations four and five are related to cognitive load theory (see Section 8.2). This theory addresses the issue of how to make it easier for learners or computer users to make sense of

the material they encounter. In this context, Sweller et al. [2011] have formulated several recommendations. We present one exemplary recommendation especially relevant for the design
of visualizations of multilayer networks. Huang et al. [2009] also address the problem of how to
investigate complex node-link diagrams which is related to our fifth and last recommendation.

R4. Cognitive load theory shows that it is especially demanding to relate facts that are presented in a disjoint way on the screen. Therefore, relationships between different layers in
multilayer networks should be emphasized so that cognitive load for remembering relationships while solving tasks does not become too heavy [Sweller, 1988]. It is still an open
question how such relationships could be emphasized.

R5. So far, researchers have predominantly used simple tasks to evaluate graph visualizations.
The results of such investigations are, in many cases, not valid for large and complex graph
visualizations. Complex tasks should be used more often to evaluate graph visualizations.

We think that the evaluation of multilayer networks is a challenging research area. There are still
many open issues that have to be solved, so that users can interact with such networks effortlessly
and derive valuable insights from such networks efficiently.

8.4   CHALLENGES AND OPPORTUNITIES

An overview of the literature indicates that the design of multilayer network visualizations is a
challenging issue. One of the most important problems in this context is cognitive load. Users
easily feel overwhelmed by the large amount of information presented to them in such types of
visualizations. There are several possibilities to overcome this issue that have also been discussed
for similar areas within the field of information visualization. Aggregation and interaction are
tools that especially might help to alleviate cognitive load. Nevertheless, more research is necessary to specify which kinds of aggregation and interaction are beneficial. In addition, these
recommendations are mainly derived from other areas of information visualization, and it is an
open question whether they are applicable for our special case of multilayer networks. An additional challenge is that many evaluation studies in information visualization were conducted
using simple tasks. This is understandable because such tasks allow the application of more rigorous testing methods. Nevertheless, it is not obvious that the results from such tasks generalize to
more complex tasks and open ended exploration processes that are more typical for real-life applications. Another challenge concerns the issue whether matrix representations can be usefully
adopted for visualizing multilayer networks. Research indicates that matrix representations are
especially beneficial for more dense and/or large networks. Further research can clarify whether
this is also true for multilayer networks.

Another challenge is the choice of appropriate methodologies for the investigation. It is
not clear which methodology or mix of methodologies is appropriate for studying the usability
and utility of multilayer network visualizations. One obvious choice would be to use eye-tracking
that enables researchers to identify the gaze direction of users. In the case of multilayer networks

this helps researchers to check whether the users detect relevant relationships or not. Another open issue concerns the use of laboratory studies. Interacting with multilayer networks requires extended exploration processes. It is difficult to investigate such processes in a laboratory. Field studies seem to be especially appropriate in this context. It should be mentioned, however, that field studies are labor-intensive and time-consuming. It is also possible that long-term investigations to study complex exploration processes is a beneficial approach in this context.

Task taxonomies are widely accepted to be useful for the evaluation process, see Kerracher and Kennedy [2017]; and the tasks described in Section 4.1 should support the evaluation of multilayer visualization systems and techniques. As already highlighted in the introduction of this chapter on evaluation, there is a lack of empirical evaluation for multilayer network visualizations. Crowdsourcing offers a lot of promise for information visualization [Borgo et al., 2017], particularly for evaluation. A survey on evaluation by using crowdsourcing in information visualization has shown that while the tasks for crowdsourcing-based evaluations are in the majority of the cases simple tasks [Borgo et al., 2018], more complex tasks are possible. Many existing crowdsourcing platforms do not lend themselves to tasks that are highly interactive due to technical restrictions. However, the development of new platforms driven by academic needs, such as suggested by Hirth et al. [2017], may simplify evaluating more complex tasks. Crowdsourcing may be useful to address the lack of evaluation of approaches to multilayer network visualizations, but the complexity of the tasks and the data sets, for the moment, makes it challenging.

To conclude, evaluation can help to clarify how users perceive multilayer networks. Results from psychology and HCI can be used to derive recommendations for the design of such visualizations. A challenge in this context is that it is sometime not straightforward to apply such results easily. Nevertheless, we think that some tentative recommendations can be derived from empirical research in HCI and psychology. Results indicate that aggregation and interaction are possibilities to support the sensemaking processes of users. Cognitive load theory indicates that processing many elements simultaneously is a challenge. This may, for example, lead to the split-attention effect that can be overcome by explicitly relating elements that belong to each other. Finally, we would like to point out that there are still many open issues that have to be clarified. Visualizations of multilayer networks can only be used efficiently if they are designed in a way so that human users can derive insights easily.

CHAPTER 9

Conclusions

Throughout this book we have explored the many different perspectives that are required for the successful visual analysis of multilayer networks. We have shown that multilayer network problems are at the intersection of domain and data, and that a multilayer network approach to visualization and analysis can be applied to many domains. We have discussed in detail how the current state of the art techniques for visualization, interaction, and analysis can be applied to the multilayer case (even when attributes are associated), covering both those that are conceived for the multilayer case, and interesting techniques that may be adapted. We have examined the importance of the layer being considered as an entity in and of itself, and have identified categories of tasks related to layers that are not covered by existing network task taxonomies. We have also discussed the many issues and challenges when it comes to evaluating multilayer visualization tools and techniques, an important topic as many existing evaluation approaches may not be suitable.

As stated in Chapter 1, one of our goals is to motivate those researchers and practitioners who want to push forward the boundaries of the visual analysis of multilayer networks, so we have identified research challenges and opportunities that will allow them to do so in each chapter. To further support this goal, we finish the book with summary conclusions for each chapter.

MULTILAYER NETWORKS ACROSS DOMAINS

We believe that the visualization of multilayer networks will play an important role in the future of network visualization and, by working closely with the field of complex systems and the application domains, researchers can uncover and find solutions to many new visualization related challenges. As the application of complex systems research becomes more popular, more application domains will take advantage of the ability to better model and handle the complexity inherent in the systems being studied. Bringing the visualization community closer to the application domain communities, as well as the complex systems communities, will result in improved outcomes for all involved. Organizing interdisciplinary workshops and seminars that include representatives from all communities was found to foster innovation in both science and technology, per Kostoff [1999].

Kivelä et al. [2014] discuss the range of data definitions (heterogeneous, multiplex, etc.) that are covered by their framework. Re-framing a user's problem within the multilayer framework may prevent commonalities between problems being obscured by nomenclature, but more importantly it will give the visualization researchers more exposure to application domain researchers addressing multilayer network problems.

As experts from various backgrounds get together, they will encounter new and interesting challenges and will need novel visualization (and visual analytics) approaches to address these problems. Such interdisciplinary gatherings need however to follow certain guidelines to maximize their benefit and avoid common pitfalls, as described by Brown et al. [2015]. Collaborations should not only be between those with a background in research. Visualization is an approach by which complex system problems can be more easily understood by the public and commercial entities. Fostering closer interaction with industry will provide new real world data, which in turn will expose new challenges and innovation.

THE LAYER AS AN ENTITY

We have introduced in Chapter 3 the notion of layers when used as entities. Although central to the concept of multilayer networks, approaches that focus mainly on layers still remain very scarce. Nevertheless, we have identified three key contributions when it comes to handling layers *per se*. *Donatien* [Hascoët and Dragicevic, 2012] introduces direct manipulation of layers so users may define their multilayer network at will. Interdonato et al. [2020] provide a wide panel of manipulations to simplify and create layers so they become relevant units of analysis for users. *Detangler* [Renoust et al., 2015] presents a visualization design that takes advantage of layers represented as entities, with interactions unique to multilayer networks.

Many challenges remain open for the visualization community. To begin with, following on from the recommendations of Interdonato et al. [2020], framing the low-level tasks specific to visualizing and interacting with layers as entities will give a framework to develop further layer visualizations. The modeling of layers from raw data is the key for a successful analysis of the complex problems addressed with multilayer networks. To this end, there is a lot to develop in terms of interactive definition and layer manipulations. Another challenge that remains completely open, is that of considering aspects themselves as entities, and handling aspects as part of the definition of the layers of a multilayer network. Since aspects share a profound relationship with attributes, there is much to be inspired from multivariate network visualization. Finally, the avenue is open for interaction design proper to layers, in the same fashion as *Detangler*. Such interactions will make it easier to integrate the level of abstraction of layers in a multilayer network visual analytics system.

TASK TAXONOMY FOR MULTILAYER NETWORKS

In a multilayer network, layers are basic elements of the network as much as nodes and edges. Thus, tasks associated to layers are low-level tasks. In Chapter 4, we proposed low-level tasks centered on layers. As the usage of multilayer networks increases, this taxonomy should evolve based on new usages and interactions. Our taxonomy may still need to be extended, when new multilayer network visualization challenges are encountered. For instance, defining an initial layering may be seen as a category itself, and not as an extension of task categories **C1/C2**.

Consolidating and refining multilayer network tasks with the typology of Brehmer and Munzner [2013], and developing higher-level task descriptions with the domains will allow for a better understanding of both the core elements of problems across domains and the full range of solutions available.

VISUALIZATION OF NODES AND RELATIONSHIPS ACROSS LAYERS

Many of the techniques discussed in Chapter 5 are established visualization techniques which can be applied to the multilayer network case. It follows naturally that many of the current research opportunities that apply to network visualization in general are of interest to those who are visualizing multilayer network data. Hybrid visualization combining multiple visualization approaches may be useful for visualizing layers with different structural properties. Immersive visualization also offers many research opportunities for information visualization researchers in general. The potential benefits of immersive visualization may be even more relevant to the multilayer case given the increased complexity of the data and the extra space afforded by a third dimension to the visualization of separate layers.

Understanding the relationships between layers is an important task (Task category **A**). This can be supported through interaction as done by Renoust et al. [2015]. However, from a pure visualization perspective edge routing, combined with an appropriate node layout is the most promising approach, if the layers are not being visualized in a summary fashion.

INTERACTING WITH AND ANALYZING MULTILAYER NETWORKS

Unsurprisingly, many of the interaction opportunities for future work related to multilayer networks focus on the layer. Existing approaches for defining and manipulating layers are quite limited, leaving much room for novel research. Interaction as a means of comparing, analyzing, and understanding layers and how they relate to each other is also an area that has much scope for further research. However, the most pertinent research direction for the visual analytics of multilayer network relates to leveraging the many multilayer specific metrics. Plenty of centralities and analytic techniques are available in the literature, that should be inspiring new visualization applications and techniques. These techniques may provide new insight to end-users and help provide them answers to the wide range of questions that arise across the myriad of application domains.

ATTRIBUTE VISUALIZATION AND MULTILAYER NETWORKS

Layers in multilayer networks add another dimension of complexity to the challenge of multivariate network visualization. Future research opportunities for multilayer network visualization with respect to the visual analysis of associated attributes can be grouped into four high-level challenges.

The first challenge concerns mixed attributes. Real-world data sets are often composed of various attribute types that are mixed together and should be analyzed as a whole. Especially for multilayer networks, the interactive visualization of this attribute setting is highly challenging. The next challenge centers on the use of dedicated network centralities. Classical network centrality measurements have already been adapted for multilayer networks. But there is a need of development for novel visualizations that consider such measurements in their visual design, e.g., for supporting cross-layer comparisons (Task category **B**). Another challenge concerns taking advantage of alternative representations. Visual network representations going beyond traditional graph layouts, such as attribute-driven layouts, have not been considered to date and should be adapted and/or enhanced for the multilayer case. Finally, recent works in the area of network embeddings have not significantly considered multilayer networks so far. The development of embedding algorithms that consider multiple layers could be a very promising direction of future research.

EVALUATION OF MULTILAYER NETWORK VISUALIZATION SYSTEMS AND TECHNIQUES

Evaluation can help to clarify how users perceive multilayer network visualization. Results from psychology and HCI can be used to derive recommendations for the design of such visualizations. A challenge in this context is that it is sometimes not straightforward to apply such results easily. Nevertheless, we think that some tentative recommendations can be derived from empirical research in HCI and psychology. Results indicate that aggregation and interaction are possibilities to support the sensemaking processes of users. Cognitive load theory indicates that processing many elements simultaneously is a challenge. This may, for example, lead to the split-attention effect that can be overcome by explicitly relating elements that belong to each other. Finally, we would like to point out that there are still many open issues that have to be clarified. Multilayer network visual analytics tools can only be used efficiently if they are designed in a way so that human users can derive insights easily.

Bibliography

J. Abello, F. Van Ham, and N. Krishnan. Ask-GraphView: A large scale graph visualization system. *IEEE Transactions on Visualization and Computer Graphics*, 12(5):669–676, 2006. DOI: 10.1109/tvcg.2006.120 91

J. Abello, S. Hadlak, H. Schumann, and H.-J. Schulz. A modular degree-of-interest specification for the visual analysis of large dynamic networks. *IEEE Transactions on Visualization and Computer Graphics*, 20(3):337–350, 2013. DOI: 10.1109/tvcg.2013.109 83

S. Agarwal, A. Tomar, and J. Sreevalsan-Nair. Nodetrix-Multiplex: Visual analytics of multiplex small world networks. In H. Cherifi, S. Gaito, W. Quattrociocchi, and A. Sala, Eds., *Complex Networks and their Applications V*, vol. 693 of *Studies in Computational Intelligence*, pages 579–591, Springer, Milano, Italy, 2017. DOI: 10.1007/978-3-319-50901-3_46 31

V. Aguiar-Pulido, W. Huang, V. Suarez-Ulloa, T. Cickovski, K. Mathee, and G. Narasimhan. Metagenomics, metatranscriptomics, and metabolomics approaches for microbiome analysis: Supplementary issue: Bioinformatics methods and applications for big metagenomics data. *Evolutionary Bioinformatics*, 12s1:EBO.S36436, 2016. DOI: 10.4137/ebo.s36436 17

J. Ahn, C. Plaisant, and B. Shneiderman. A task taxonomy for network evolution analysis. *IEEE Transactions on Visualization and Computer Graphics*, 99(PP):365–376, 2013. DOI: 10.1109/tvcg.2013.238 37, 41, 42

C. G. Akcora, Y. R. Gel, and M. Kantarcioglu. Blockchain: A graph primer. *CoRR*, 2017. http://arxiv.org/abs/1708.08749 18

A. Aleta and Y. Moreno. Multilayer networks in a nutshell. *Annual Review of Condensed Matter Physics*, 10(1):45–62, 2019. DOI: 10.1146/annurev-conmatphys-031218-013259 11, 12, 28

B. Alper, B. Bach, N. Henry Riche, T. Isenberg, and J.-D. Fekete. Weighted graph comparison techniques for brain connectivity analysis. In *Proc. of the SIGCHI Conference on Human Factors in Computing Systems, CHI'13*, pages 483–492, ACM, Paris, France, 2013. DOI: 10.1145/2470654.2470724 48

B. Alsallakh, W. Aigner, S. Miksch, and H. Hauser. Radial sets: Interactive visual analysis of large overlapping sets. *IEEE Transactions on Visualization and Computer Graphics*, 19(12):2496–2505, 2013. DOI: 10.1109/tvcg.2013.184 52

B. Alsallakh, L. Micallef, W. Aigner, H. Hauser, S. Miksch, and P. Rodgers. The state-of-the-art of set visualization. *Computer Graphics Forum*, 35(1):234–260, 2016. DOI: 10.1111/cgf.12722 64

R. Amar, J. Eagan, and J. Stasko. Low-level components of analytic activity in information visualization. In *IEEE Symposium on Information Visualization (INFOVIS)*, pages 111–117, 2005. DOI: 10.1109/INFVIS.2005.1532136 66

R. Amato, N. E. Kouvaris, M. San Miguel, and A. Díaz-Guilera. Opinion competition dynamics on multiplex networks. *New Journal of Physics*, 19(123019), 2017. DOI: 10.1088/1367-2630/aa936a 72

K. Andrews, M. Wohlfahrt, and G. Wurzinger. Visual graph comparison. In *2009 13th International Conference Information Visualisation (IV)*, 00:62–67, July 2009. DOI: 10.1109/iv.2009.108 56

N. Andrienko and G. Andrienko. *Exploratory Analysis of Spatial and Temporal Data: A Systematic Approach.* Springer, Berlin, Heidelberg, 2006. DOI: 10.1007/3-540-31190-4 42

L. Angori, W. Didimo, F. Montecchiani, D. Pagliuca, and A. Tappini. Hybrid graph visualizations with ChordLink: Algorithms, experiments, and applications. *IEEE Transactions on Visualization and Computer Graphics*, 2020. DOI: 10.1109/tvcg.2020.3016055 52, 62, 133

D. Archambault, T. Munzner, and D. Auber. Topolayout: Multilevel graph layout by topological features. *IEEE Transactions on Visualization and Computer Graphics*, 13(2):305–317, March 2007. DOI: 10.1109/tvcg.2007.46 53

D. Archambault, T. Munzner, and D. Auber. Grouseflocks: Steerable exploration of graph hierarchy space. *IEEE Transactions on Visualization and Computer Graphics*, 14(4):900–913, July 2008. DOI: 10.1109/tvcg.2008.34 91

D. Archambault, H. C. Purchase, and B. Pinaud. The readability of path-preserving clusterings of graphs. *Computer Graphics Forum*, 29(3):1173–1182, 2010. DOI: 10.1111/j.1467-8659.2009.01683.x 90

D. Archambault, T. Munzner, and D. Auber. Tugging graphs faster: Efficiently modifying path-preserving hierarchies for browsing paths. *IEEE Transactions on Visualization and Computer Graphics*, 17(3):276–289, March 2011. DOI: 10.1109/tvcg.2010.60 91

D. Archambault, J. Abello, J. Kennedy, S. Kobourov, K.-L. Ma, S. Miksch, C. Muelder, and A. C. Telea. *Temporal Multivariate Networks*, pages 151–174. Springer International Publishing, Cham, 2014. DOI: 10.1007/978-3-319-06793-3_8 81

M. Atzmueller, S. Doerfel, A. Hotho, F. Mitzlaff, and G. Stumme. Face-to-face contacts at a conference: dynamics of communities and roles. In M. Atzmueller, A. Chin, D. Helic, and A. Hotho, Eds., *Modeling and Mining Ubiquitous Social Media (MSM)*, vol. 7472 of *Lecture Notes in Computer Science*, pages 21–39, Springer Berlin Heidelberg, Athens, Greece, 2012. DOI: 10.1007/978-3-642-33684-3_2 26

B. Bach, N. Henry-Riche, T. Dwyer, T. Madhyastha, J.-D. Fekete, and T. Grabowski. Small MultiPiles: Piling time to explore temporal patterns in dynamic networks. *Computer Graphics Forum*, 2015. DOI: 10.1111/cgf.12615 81

B. Bach, R. Sicat, J. Beyer, M. Cordeil, and H. Pfister. The hologram in my hand: How effective is interactive exploration of 3D visualizations in immersive tangible augmented reality? *IEEE Transactions on Visualization and Computer Graphics*, 24(1):457–467, 2018. DOI: 10.1109/tvcg.2017.2745941 63

M. Balzer and O. Deussen. Level-of-detail visualization of clustered graph layouts. In *6th International Asia-Pacific Symposium on Visualization*, pages 133–140, 2007. DOI: 10.1109/apvis.2007.329288 63

L. Bargigli, G. Di Iasio, L. Infante, F. Lillo, and F. Pierobon. The multiplex structure of interbank networks. *Quantitative Finance*, 15(4):673–691, 2015. DOI: 10.1080/14697688.2014.968356 18

F. Battiston, V. Nicosia, and V. Latora. Structural measures for multiplex networks. *Physical Review E*, 89(3):032804, 2014. DOI: 10.1103/physreve.89.032804 28, 71, 80

M. Bazzi, M. A. Porter, S. Williams, M. McDonald, D. J. Fenn, and S. D. Howison. Community detection in temporal multilayer networks, with an application to correlation networks. *Multiscale Modeling and Simulation*, 14(1):1–41, 2016. DOI: 10.1137/15m1009615 18

F. Beck, M. Burch, S. Diehl, and D. Weiskopf. A taxonomy and survey of dynamic graph visualization. *Computer Graphics Forum*, 36(1):133–159, 2017. DOI: 10.1111/cgf.12791 16, 37, 41, 44, 81

B. Bentley, R. Branicky, C. L. Barnes, Y. L. Chew, E. Yemini, E. T. Bullmore, P. E. Vértes, and W. R. Schafer. The multilayer connectome of caenorhabditis elegans. *PLOS Computational Biology*, 12(12):1–31, 2016. DOI: 10.1371/journal.pcbi.1005283 53

P. Berger, H. Schumann, and C. Tominski. Visually exploring relations between structure and attributes in multivariate graphs. In *23rd International Conference Information Visualisation (IV)*, pages 261–268, IEEE, 2019. DOI: 10.1109/iv.2019.00051 82

I. Bertazzi, S. Huet, G. Deffuant, and F. Gargiulo. The anatomy of a Web of trust: The bitcoin-OTC market. In *Social Informatics*, pages 228–241, Springer, 2018. DOI: 10.1007/978-3-030-01129-1_14 19

A. Bezerianos, F. Chevalier, P. Dragicevic, N. Elmqvist, and J.-D. Fekete. GraphDice: A system for exploring multivariate social networks. *Computer Graphics Forum*, 29(3):863–872, 2010. DOI: 10.1111/j.1467-8659.2009.01687.x 29, 38, 41

G. Bianconi. Multilayer networks: Dangerous liaisons? *Nature Physics*, 10(10):712–714, October 2014. DOI: 10.1038/nphys3097 7

A. Bigelow, C. Nobre, M. Meyer, and A. Lex. Origraph: Interactive network wrangling. In *IEEE Conference on Visual Analytics Science and Technology (VAST)*, pages 81–92, arXiv:1812.06337, 2019. DOI: 10.1109/vast47406.2019.8986909 29, 33, 73, 74

A. E. Biondo, A. Pluchino, and A. Rapisarda. Informative contagion dynamics in a multilayer network model of financial markets. *Italian Economic Journal*, 3(3):343–366, 2017. DOI: 10.1007/s40797-017-0052-4 18

S. Boccaletti, G. Bianconi, R. Criado, C. I. Del Genio, J. Gómez-Gardenes, M. Romance, I. Sendina-Nadal, Z. Wang, and M. Zanin. The structure and dynamics of multilayer networks. *Physics Reports*, 544(1):1–122, 2014. DOI: 10.1016/j.physrep.2014.07.001 16, 64

R. Bookstaber and D. Kenett. Looking deeper, seeing more: A multilayer map of the financial system. Briefs 16-06, Office of Financial Research, U.S. Department of the Treasury, 2016. https://EconPapers.repec.org/RePEc:ofr:briefs:16--06 18

S. P. Borgatti and M. G. Everett. Network analysis of 2-mode data. *Social Networks*, 19(3):243–269, 1997. DOI: 10.1016/s0378-8733(96)00301-2 13, 14, 71

S. P. Borgatti, A. Mehra, D. J. Brass, and G. Labianca. Network analysis in the social sciences. *Science*, 323(5916):892–895, 2009. DOI: 10.1126/science.1165821 12, 18, 27

R. Borgo, B. Lee, B. Bach, S. Fabrikant, R. Jianu, A. Kerren, S. Kobourov, F. McGee, L. Micallef, T. von Landesberger, et al. Crowdsourcing for information visualization: Promises and pitfalls. In *Evaluation in the Crowd. Crowdsourcing and Human-Centered Experiments*, vol. 10264 of *Lecture Notes in Computer Science*, pages 96–138, Springer, 2017. DOI: 10.1007/978-3-319-66435-4_5 95

R. Borgo, L. Micallef, B. Bach, F. McGee, and B. Lee. Information visualization evaluation using crowdsourcing. *Computer Graphics Forum*, 37(3):573–595, 2018. DOI: 10.1111/cgf.13444 95

G. Bothorel, M. Serrurier, and C. Hurter. Visualization of frequent itemsets with nested circular layout and bundling algorithm. In *International Symposium on Visual Computing*, vol. 8034 of *Lecture Notes in Computer Science*, pages 396–405, Springer, Rethymnon, Crete, Greece, 2013. DOI: 10.1007/978-3-642-41939-3_38 51

R. Bourqui, D. Ienco, A. Sallaberry, and P. Poncelet. Multilayer graph edge bundling. In *IEEE Pacific Visualization Symposium (PacificVis)*, pages 184–188, Taipei, Taiwan, April 2016. DOI: 10.1109/pacificvis.2016.7465267 38, 45, 58, 59, 63, 134

M. Brehmer and T. Munzner. A multi-level typology of abstract visualization tasks. *IEEE Transactions on Visualization and Computer Graphics*, 19(12):2376–2385, 2013. DOI: 10.1109/tvcg.2013.124 7, 40, 42, 44, 99

R. L. Breiger. The duality of persons and groups. *Social Forces*, 53(2):181–190, 1974. DOI: 10.2307/2576011 14

D. A. Bright, C. Greenhill, A. Ritter, and C. Morselli. Networks within networks: Using multiple link types to examine network structure and identify key actors in a drug trafficking operation. *Global Crime*, 16(3):219–237, 2015. DOI: 10.1080/17440572.2015.1039164 6

P. Bródka, A. Chmiel, M. Magnani, and G. Ragozini. Quantifying layer similarity in multiplex networks: A systematic study. *Royal Society Open Science*, 5(8):171747, 2018. DOI: 10.1098/rsos.171747 28

R. R. Brown, A. Deletic, and T. H. Wong. Interdisciplinarity: How to catalyse collaboration. *Nature News*, 525(7569):315, 2015. DOI: 10.1038/525315a 98

S. V. Buldyrev, R. Parshani, G. Paul, H. E. Stanley, and S. Havlin. Catastrophic cascade of failures in interdependent networks. *Nature*, 464:1025–1028, 2010. DOI: 10.1038/nature08932 20

M. Burch and S. Diehl. TimeRadarTrees: Visualizing dynamic compound digraphs. *Computer Graphics Forum*, 27(3):823–830, 2008. DOI: 10.1111/j.1467-8659.2008.01213.x 81

M. Burch, M. Fritz, F. Beck, and S. Diehl. TimeSpiderTrees: A novel visual metaphor for dynamic compound graphs. In *Proc. of the IEEE Symposium on Visual Languages and Human-Centric Computing, VL/HCC*, pages 168–175, 2010. DOI: 10.1109/vlhcc.2010.31 82

M. Burch, M. Höferlin, and D. Weiskopf. Layered TimeRadarTrees. In *Proc. of the 15th International Conference on Information Visualisation, IV*, pages 18–25, IEEE, 2011. DOI: 10.1109/iv.2011.93 82, 134

M. Burch, W. Huang, M. Wakefield, H. C. Purchase, D. Weiskopf, and J. Hua. The state-of-the-art in empirical user evaluation of graph visualizations. *IEEE Access*, 9:4173–4198, 2021. DOI: 10.1109/access.2020.3047616 89

R. Burt and T. Schøtt. Relation content in multiple networks. *Social Science Research*, 14(4):287–308, 1985. DOI: 10.1016/0049-089X(85)90014-6 xi, 1, 2, 6, 12, 17

D. Cai, Z. Shao, X. He, X. Yan, and J. Han. Community mining from multi-relational networks. In A. M. Jorge, L. Torgo, P. Brazdil, R. Camacho, and J. Gama, Eds., *Knowledge Discovery in Databases: PKDD 2005*, pages 445–452, Springer, Berlin, Heidelberg, 2005. DOI:
10.1007/11564126_44 6

G. Caldarelli. Foundational research on multilevel complex networks and systems, 2012. https:
//cordis.europa.eu/project/id/317532 12

N. Cao, J. Sun, Y.-R. Lin, D. Gotz, S. Liu, and H. Qu. FacetAtlas: Multifaceted visualization
for rich text corpora. *IEEE Transactions on Visualization and Computer Graphics*, 16(6):1172–
1181, 2010. DOI: 10.1109/tvcg.2010.154 15, 29, 38, 40

N. Cao, Y.-R. Lin, L. Li, and H. Tong. G-Miner: Interactive visual group mining on multivariate graphs. In *Proc. of the 33rd Annual ACM Conference on Human Factors in Computing
Systems*, pages 279–288, Seoul, Republic of Korea, 2015. DOI: 10.1145/2702123.2702446
38, 39, 40

A. Cardillo, J. Gómez-Gardeñes, M. Zanin, M. Romance, D. Papo, F. del Pozo, and S. Boccaletti. Emergence of network features from multiplexity. *Nature*, 3:1344, February 2013.
DOI: 10.1038/srep01344 6

C. Cetinkaya and E. W. Knightly. Opportunistic traffic scheduling over multiple network paths.
In *IEEE INFOCOM*, 3:1928–1937, 2004. DOI: 10.1109/infcom.2004.1354602 20

C.-H. Chen. Generalized association plots: Information visualization via iteratively generated correlation matrices. *Statistica Sinica*, 12(1):7–29, 2002. http://www.jstor.org/stable/
24307033 47

J. Chuang, C. D. Manning, and J. Heer. Termite: Visualization techniques for assessing textual
topic models. In *Proc. of the International Working Conference on Advanced Visual Interfaces,
AVT'12*, pages 74–77, ACM, Capri Island, Italy, 2012. DOI: 10.1145/2254556.2254572 47

C. Collins and S. Carpendale.   Vislink: Revealing relationships amongst visualizations.
IEEE Transactions on Visualization and Computer Graphics, 13(6):1192–1199, 2007. DOI:
10.1109/tvcg.2007.70521 38, 62

C. Collins, G. Penn, and S. Carpendale. Bubble sets: Revealing set relations with isocon-
tours over existing visualizations. *IEEE Transactions on Visualization and Computer Graphics*,
15(6):1009–1016, 2009. DOI: 10.1109/tvcg.2009.122 63

A. Cottica, A. Hassoun, M. Manca, J. Vallet, and G. Melançon. Semantic social networks: A
mixed methods approach to digital ethnography. *Field Methods*, 32(3):274–290, 2020. DOI:
10.1177/1525822x20908236 38, 40

L. Cottret, D. Wildridge, F. Vinson, M. P. Barrett, H. Charles, M.-F. Sagot, and F. Jourdan. Metexplore: A web server to link metabolomic experiments and genome-scale metabolic networks. *Nucleic Acids Research*, 38(Web server issue):W132–W137, 2010. DOI: 10.1093/nar/gkq312 17

H. K. Crabb, J. L. Allen, J. M. Devlin, S. M. Firestone, M. A. Stevenson, and J. R. Gilkerson. The use of social network analysis to examine the transmission of salmonella spp. within a vertically integrated broiler enterprise. *Food Microbiology*, 71:73–81, 2017. DOI: 10.1016/j.fm.2017.03.008 18

T. Crnovrsanin, C. W. Muelder, R. Faris, D. Felmlee, and K.-L. Ma. Visualization techniques for categorical analysis of social networks with multiple edge sets. *Social Networks*, 37(Supplement C):56–64, 2014. DOI: 10.1016/j.socnet.2013.12.002 5, 6, 18, 49, 52, 59, 60, 63, 89, 134

E. Cuenca, A. Sallaberry, D. Ienco, and P. Poncelet. Visual querying of large multilayer graphs. In *Proc. of the 30th International Conference on Scientific and Statistical Database Management, SSDBM'18*, Association for Computing Machinery, 2018. DOI: 10.1145/3221269.3223027 73

E. Cuenca, A. Sallaberry, D. Ienco, and P. Poncelet. Vertigo: A visual platform for querying and exploring large multilayer networks. *IEEE Transactions on Visualization and Computer Graphics*, pages 1–1, 2021. DOI: 10.1109/tvcg.2021.3067820 73

P. Cui, X. Wang, J. Pei, and W. Zhu. A survey on network embedding. *IEEE Transactions on Knowledge and Data Engineering*, 31(5):833–852, 2019. DOI: 10.1109/TKDE.2018.2849727 85

L. de Carvalho, G. Borelli, A. Camargo, M. de Assis, S. de Ferraz, M. Fiamenghi, J. José, L. Mofatto, S. Nagamatsu, G. Persinoti, N. Silva, A. Vasconcelos, G. Pereira, and M. Carazzolle. Bioinformatics applied to biotechnology: A review towards bioenergy research. *Biomass and Bioenergy*, 123:195–224, 2019. DOI: 10.1016/j.biombioe.2019.02.016 17

M. De Domenico. Multilayer modeling and analysis of human brain networks. *GigaScience*, 6(5):1–8, 2017. DOI: 10.1093/gigascience/gix004 53

M. De Domenico. Multilayer network modeling of integrated biological systems: Comment on network science of biological systems at different scales: A review by Gosak et al. *Physics of Life Reviews*, 24:149–152, 2018. DOI: 10.1016/j.plrev.2017.12.006 7

M. De Domenico, A. Solé-Ribalta, E. Cozzo, M. Kivelä, Y. Moreno, M. A. Porter, S. Gómez, and A. Arenas. Mathematical formulation of multilayer networks. *Phys. Rev. X*, 3:041022, December 2013. DOI: 10.1103/physrevx.3.041022 25, 71

M. De Domenico, M. A. Porter, and A. Arenas. MuxViz: A tool for multilayer analysis and visualization of networks. *Journal of Complex Networks*, 3(2):159–176, 2015. DOI: 10.1093/comnet/cnu038 8, 28, 31, 34, 38, 45, 47, 51, 53, 54, 72, 83, 84, 134

J. V. L. de Jeude, T. Aste, and G. Caldarelli. The multilayer structure of corporate networks. *New Journal of Physics*, 21(2):025002, 2019. DOI: 10.1088/1367-2630/ab022d 18

C. M. Delude. Deep phenotyping: The details of disease. *Nature*, 527(7576):S14, 2015. DOI: 10.1038/527s14a 17

S. Derrible. Complexity in future cities: The rise of networked infrastructure. *International Journal of Urban Sciences*, 21(sup1):68–86, 2017. DOI: 10.1080/12265934.2016.1233075 6, 20

G. Di Battista, P. Eades, R. Tamassia, and I. G. Tollis. *Graph Drawing: Algorithms for the Visualization of Graphs*. Prentice Hall, 1999. 88

E. Di Giacomo, W. Didimo, G. Liotta, and P. Palladino. Visual analysis of one-to-many matched graphs. In I. G. Tollis and M. Patrignani, Eds., *Graph Drawing*, pages 133–144, Springer Berlin Heidelberg, Berlin, Heidelberg, 2009. DOI: 10.1007/978-3-642-00219-9_14 56

M. E. Dickison, M. Magnani, and L. Rossi. *Multilayer Social Networks*. Cambridge University Press, 2016a. DOI: 10.1017/cbo9781139941907 6, 8, 72, 78

M. E. Dickison, M. Magnani, and L. Rossi. *Visualizing Multilayer Networks*, pages 79–95. Cambridge University Press, 2016b. DOI: 10.1017/cbo9781139941907.005 28

K. Dinkla, M. J. Van Kreveld, B. Speckmann, and M. A. Westenberg. Kelp diagrams: Point set membership visualization. *Computer Graphics Forum*, 31(3pt1):875–884, 2012. DOI: 10.1111/j.1467-8659.2012.03080.x 73

M. Domenico, A. Sol-Ribalta, E. Omodei, S. Gmez, and A. Arenas. Centrality in interconnected multilayer networks. *Nature Communications*, 6(6868):2013. DOI: 10.1038/ncomms7868 83

M. M. Dow. Galton's problem as multiple network autocorrelation effects: Cultural trait transmission and ecological constraint. *Cross-Cultural Research*, 41(4):336–363, 2007. DOI: 10.1177/1069397107305452 18

C. Ducruet. Multilayer dynamics of complex spatial networks: The case of global maritime flows (1977–2008). *Journal of Transport Geography*, 60:47–58, 2017. DOI: 10.1016/j.jtrangeo.2017.02.007 6, 33, 46, 54, 55, 57, 134

R. Dunbar, V. Arnaboldi, M. Conti, and A. Passarella. The structure of online social networks mirrors those in the offline world. *Social Networks*, 43:39–47, 2015. DOI: 10.1016/j.socnet.2015.04.005 2, 31

C. Dunne and B. Shneiderman. Motif simplification: Improving network visualization readability with fan, connector, and clique glyphs. In *Proc. of the SIGCHI Conference on Human Factors in Computing Systems, CHI'13*, pages 3247–3256, ACM, 2013. DOI: 10.1145/2470654.2466444 90

C. Dunne, N. Henry Riche, B. Lee, R. Metoyer, and G. Robertson. Graphtrail: Analyzing large multivariate, heterogeneous networks while supporting exploration history. In *Proc. of the SIGCHI Conference on Human Factors in Computing Systems, CHI'12*, pages 1663–1672, ACM, Austin, Texas, 2012. DOI: 10.1145/2207676.2208293 6, 19, 38, 56, 80, 88

P. Eichmann, D. Edge, N. Evans, B. Lee, M. Brehmer, and C. White. Orchard: Exploring multivariate heterogeneous networks on mobile phones. *Computer Graphics Forum*, 39(3):115–126, 2020. DOI: 10.1111/cgf.13967 50, 69, 80

S. Engle and S. Whalen. Visualizing distributed memory computations with hive plots. In *Proc. of the 9th International Symposium on Visualization for Cyber Security (VizSec'12)*, pages 56–63, ACM, Seattle, WA, 2012. DOI: 10.1145/2379690.2379698 50, 51

G. Fagherazzi. Deep digital phenotyping and digital twins for precision health: Time to dig deeper. *Journal of Medical Internet Research*, 22(3):e16770, March 2020. DOI: 10.2196/16770 17

J.-D. Fekete. Reorder.js: A javascript library to reorder tables and networks. In *VIS Poster*, IEEE, Chicago, 2015. https://hal.inria.fr/hal-01214274 47

S. Franconeri. The nature and status of visual resources. In D. Reisberg, Ed., *The Oxford Handbook of Cognitive Psychology*, Oxford Library of Psychology, Oxford University Press, 2013. DOI: 10.1093/oxfordhb/9780195376746.013.0010 93

M. Freire, C. Plaisant, B. Shneiderman, and J. Golbeck. ManyNets: An interface for multiple network analysis and visualization. In *Proc. of the SIGCHI Conference on Human Factors in Computing Systems, CHI'10*, pages 213–222, ACM, Atlanta, Georgia, 2010. DOI: 10.1145/1753326.1753358 18, 31, 32, 38, 83, 133

D. C. Fung, S.-H. Hong, D. Koschützki, F. Schreiber, and K. Xu. Visual analysis of overlapping biological networks. In *13th International Conference Information Visualisation*, pages 337–342, IEEE, Barcelona, Spain, 2009. DOI: 10.1109/iv.2009.55 46, 47, 54, 56

G. Gallo, G. Longo, S. Pallottino, and S. Nguyen. Directed hypergraphs and applications. *Discrete Applied Mathematics*, 42(2–3):177–201, 1993. DOI: 10.1016/0166-218x(93)90045-p 14

R. Gallotti and M. Barthelemy. The multilayer temporal network of public transport in Great Britain. *Nature Scientific Data*, 2(140056), 2015. DOI: 10.1038/sdata.2014.56 5, 20, 24, 31, 33, 53, 64

E. R. Gansner, Y. Hu, and S. Kobourov. GMap: Visualizing graphs and clusters as maps. In *IEEE Pacific Visualization Symposium (PacificVis)*, pages 201–208, 2010. DOI: 10.1109/pacificvis.2010.5429590 63

J. Gao, S. V. Buldyrev, H. E. Stanley, and S. Havlin. Networks formed from interdependent networks. *Nature Physics*, 8(1):40–48, 2012. DOI: 10.1038/nphys2180 1, 6

N. Garcia, B. Renoust, and Y. Nakashima. ContextNet: Representation and exploration for painting classification and retrieval in context. *International Journal of Multimedia Information Retrieval*, 9(1):17–30, 2020. DOI: 10.1007/s13735-019-00189-4 19

N. Geard and S. Bullock. Milieu and function: Toward a multilayer framework for understanding social networks. In *Workshop Proceedings of the 9th European Conference on Artificial Life (ECAL): The Emergence of Social Behaviour*, pages 1–11, Portugal, 2007. http://eprints.soton.ac.uk/id/eprint/264340 2, 6

N. Gehlenborg, S. I. O'donoghue, N. S. Baliga, A. Goesmann, M. A. Hibbs, H. Kitano, O. Kohlbacher, H. Neuweger, R. Schneider, D. Tenenbaum, et al. Visualization of omics data for systems biology. *Nature Methods*, 7(3 Suppl.):S56–68, 2010. DOI: 10.1038/nmeth.1436 17

F. Geier, W. Barfuss, M. Wiedermann, J. Kurths, and J. F. Donges. The physics of governance networks: Critical transitions in contagion dynamics on multilayer adaptive networks with application to the sustainable use of renewable resources. *The European Physical Journal Special Topics*, 228(11):2357–2369, 2019. DOI: 10.1140/epjst/e2019-900120-4 19

S. Ghani, B. C. Kwon, S. Lee, J. S. Yi, and N. Elmqvist. Visual analytics for multimodal social network analysis: A design study with social scientists. *Transactions on Visualization and Computer Graphics*, 19(12):2032–2041, 2013. DOI: 10.1109/tvcg.2013.223 6, 13, 18, 26, 38, 39, 49, 50, 56, 58, 72, 75, 88, 133

S. Ghariblou, M. Salehi, M. Magnani, and M. Jalili. Shortest paths in multiplex networks. *Scientific Reports*, 7(1):2142, 2017. DOI: 10.1038/s41598-017-01655-x 39

M. Ghoniem, H. Cambazard, J.-D. Fekete, and N. Jussien. Peeking in solver strategies using explanations visualization of dynamic graphs for constraint programming. In *Proc. of the ACM symposium on Software Visualization (SoftVis)*, pages 27–36, St. Louis, MO, 2005a. DOI: 10.1145/1056018.1056022 47

M. Ghoniem, J.-D. Fekete, and P. Castagliola. On the readability of graphs using node-link and matrix-based representations: A controlled experiment and statistical analysis. *Information Visualization*, 4(2):114–135, 2005b. DOI: 10.1057/palgrave.ivs.9500092 46, 61, 91

M. Gleicher, D. Albers, R. Walker, I. Jusufi, C. D. Hansen, and J. C. Roberts. Visual comparison for information visualization. *Information Visualization*, 10(4):289–309, 2011. DOI: 10.1177/1473871611416549 32

M. Gosak, R. Marković, J. Dolenšek, M. S. Rupnik, M. Marhl, A. Stožer, and M. Perc. Network science of biological systems at different scales: A review. *Physics of Life Reviews*, 2017. DOI: 10.1016/j.plrev.2017.11.003 6, 17

R. V. Gould. Multiple networks and mobilization in the Paris commune, 1871. *American Sociological Review*, pages 716–729, 1991. DOI: 10.2307/2096251 17

P. Goyal and E. Ferrara. Graph embedding techniques, applications, and performance: A survey. *Knowledge-Based Systems*, 151:78–94, 2018. DOI: 10.1016/j.knosys.2018.03.022 85

N. Greffard, F. Picarougne, and P. Kuntz. Visual community detection: An evaluation of 2D, 3D perspective and 3D stereoscopic displays. In M. Van Kreveld and B. Speckmann, Eds., *Graph Drawing*, pages 215–225, Springer Berlin Heidelberg, Redmond, WA, 2012. DOI: 10.1007/978-3-642-25878-7_21 57

S. Grottel, J. Heinrich, D. Weiskopf, and S. Gumhold. Visual analysis of trajectories in multi-dimensional state spaces. *Computer Graphics Forum*, 33(6):310–321, 2014. DOI: 10.1111/cgf.12352 78, 79, 134

A. Grover and J. Leskovec. Node2vec: Scalable feature learning for networks. In *Proc. of the 22nd ACM SIGKDD International Conference on Knowledge Discovery and Data Mining*, pages 855–864, Association for Computing Machinery, 2016. DOI: 10.1145/2939672.2939754 85

S. Hadlak, H. Schumann, and H.-J. Schulz. A survey of multi-faceted graph visualization. In *Eurographics Conference on Visualization (EuroVis). The Eurographics Association*, pages 1–20, Cagliary, Italy, 2015. DOI: 10.2312/eurovisstar.20151109 15, 31, 44

G. Halin, D. Hanser, and S. Kubicki. Towards an integration of the cooperative design context in collaborative tools. In *Conférence EMISA, Informationssysteme*, pages 1–12, France, 2004. https://halshs.archives-ouvertes.fr/halshs-00267836 20

A. Halu, S. Mukherjee, and G. Bianconi. Emergence of overlap in ensembles of spatial multiplexes and statistical mechanics of spatial interacting network ensembles. *Physical Review E*, 89:012806, January 2014. DOI: 10.1103/physreve.89.012806 5, 20

M. Hascoët and P. Dragicevic. Interactive graph matching and visual comparison of graphs and clustered graphs. In *Proc. of the International Working Conference on Advanced Visual Interfaces, AVI'12*, pages 522–529, ACM, Capri Island, Italy, 2012. DOI: 10.1145/2254556.2254654 5, 28, 38, 39, 40, 55, 56, 66, 67, 68, 69, 70, 72, 98, 134

L. S. Heath and A. A. Sioson. Multimodal networks: Structure and operations. *IEEE/ACM Transactions on Computational Biology and Bioinformatics (TCBB)*, 6(2):321–332, 2009. DOI: 10.1109/tcbb.2007.70243 6

J. Heer and A. Perer. Orion: A system for modeling, transformation and visualization of multi-dimensional heterogeneous networks. *Information Visualization*, 13(2):111–133, 2014. DOI: 10.1177/1473871612462152 29, 33

M. Hegarty. The cognitive science of visual-spatial displays: Implications for design. *Topics in Cognitive Science*, 3(3):446–474, 2011. DOI: 10.1111/j.1756-8765.2011.01150.x 89, 92

N. Henry, J. D. Fekete, and M. J. McGuffin. Nodetrix: A hybrid visualization of social networks. *IEEE Transactions on Visualization and Computer Graphics*, 13(6):1302–1309, 2007. DOI: 10.1109/tvcg.2007.70582 62

I. Herman, M. S. Marshall, and G. Melançon. Graph visualisation and navigation in information visualisation: A survey. *IEEE Transactions on Visualization and Computer Graphics*, 6(1):24–43, 2000. DOI: 10.1109/2945.841119 88

M. Hirth, J. Jacques, P. Rodgers, O. Scekic, and M. Wybrow. Crowdsourcing technology to support academic research. In D. Archambault, H. Purchase, and T. Hoßfeld, Eds., *Evaluation in the Crowd. Crowdsourcing and Human-Centered Experiments*, vol. 10264 of *Lecture Notes in Computer Science*, pages 70–95. Springer International Publishing, 2017. DOI: 10.1007/978-3-319-66435-4_4 95

J. Hoffswell, A. Borning, and J. Heer. Setcola: High-level constraints for graph layout. *Computer Graphics Forum*, 37(3):537–548, 2018. DOI: 10.1111/cgf.13440 53

D. Holten. Hierarchical edge bundles: Visualization of adjacency relations in hierarchical data. *IEEE Transactions on Visualization and Computer Graphics*, 12(5):741–748, September 2006. DOI: 10.1109/tvcg.2006.147 51, 57, 58, 63, 80

D. Holten and J. J. Van Wijk. Visual comparison of hierarchically organized data. *Computer Graphics Forum*, 27(3):759–766, 2008. DOI: 10.1111/j.1467-8659.2008.01205.x 38, 63, 80

D. Holten and J. J. Van Wijk. Force-directed edge bundling for graph visualization. *Computer Graphics Forum*, 28(3):983–990, 2009. DOI: 10.1111/j.1467-8659.2009.01450.x 63

R. Howard and E. Petersen. Monitoring communication in partnering projects. *Journal of Information Technology in Construction (ITCon)*, 6(1):1–16, 2002. https://www.itcon.org/2001/1 20

D. Huang, M. Tory, B. A. Aseniero, L. Bartram, S. Bateman, S. Carpendale, A. Tang, and R. Woodbury. Personal visualization and personal visual analytics. *Visualization and Computer Graphics, IEEE Transactions on*, 21(3):420–433, 2015. DOI: 10.1109/TVCG.2014.2359887 31

W. Huang, P. Eades, and S.-H. Hong. Measuring effectiveness of graph visualizations: A cognitive load perspective. *Information Visualization*, 8(3):139–152, 2009. DOI: 10.1057/ivs.2009.10 92, 94

S. R. Humayoun, H. Ezaiza, R. AlTarawneh, and A. Ebert. Social-circles exploration through interactive multi-layered chord layout. In *Proc. of the International Working Conference on Advanced Visual Interfaces (AVI)*, pages 314–315, ACM, Bari, Italy, 2016. DOI: 10.1145/2909132.2926072 52

A. Inselberg and B. Dimsdale. Parallel coordinates: A tool for visualizing multi-dimensional geometry. In *Proc. of the 1st IEEE Conference on Visualization: Visualization*, pages 361–378, San Francisco, CA, October 1990. DOI: 10.1109/visual.1990.146402 48

R. Interdonato, M. Magnani, D. Perna, A. Tagarelli, and D. Vega. Multilayer network simplification: Approaches, models and methods. *Computer Science Review*, 36:100246, 2020. DOI: 10.1016/j.cosrev.2020.100246 27, 28, 34, 43, 74, 98

W. Javed and N. Elmqvist. Exploring the design space of composite visualization. In *IEEE Pacific Visualization Symposium (PacificVis)*, pages 1–8, Songdo, Republic of Korea, 2012. DOI: 10.1109/pacificvis.2012.6183556 61

I. Jusufi. Multivariate networks: Visualization and interaction techniques. Ph.D. Thesis, Linnaeus University, Växjö, Sweden, 2013. 84

I. Jusufi, A. Kerren, and B. Zimmer. Multivariate network exploration with JauntyNets. In *17th International Conference on Information Visualisation*, pages 19–27, London, UK, July 2013. DOI: 10.1109/iv.2013.3 85, 86, 134

S. Kairam, N. H. Riche, S. Drucker, R. Fernandez, and J. Heer. Refinery: Visual exploration of large, heterogeneous networks through associative browsing. *Computer Graphics Forum*, 34(3):301–310, 2015. DOI: 10.1111/cgf.12642 33, 34, 38, 39, 56

S. Kandel, J. Heer, C. Plaisant, J. Kennedy, F. Van Ham, N. H. Riche, C. Weaver, B. Lee, D. Brodbeck, and P. Buono. Research directions in data wrangling: Visualizations and transformations for usable and credible data. *Information Visualization*, 10(4):271–288, 2011. DOI: 10.1177/1473871611415994 33

M. Kaufmann and D. Wagner. *Drawing graphs: Methods and models.* Springer, 2003. DOI: 10.1007/3-540-44969-8 88

D. Keim, G. Andrienko, J.-D. Fekete, C. Görg, J. Kohlhammer, and G. Melançon. Visual analytics: Definition, process, and challenges. In *Information Visualization: Human-Centered Issues and Perspectives*, pages 154–175, Springer, 2008. DOI: 10.1007/978-3-540-70956-5_7 65

D. Y. Kenett, M. Perc, and S. Boccaletti. Networks of networks—An introduction. *Chaos, Solitons and Fractals*, 80:1–6, 2015. DOI: 10.1016/j.chaos.2015.03.016 1, 6

N. Kerracher and J. Kennedy. Constructing and evaluating visualisation task classifications: Process and considerations. *Computer Graphics Forum*, 36(3):47–59, 2017. DOI: 10.1111/cgf.13167 95

N. Kerracher, J. Kennedy, and K. Chalmers. The design space of temporal graph visualisation. In N. Elmqvist, M. Hlawitschka, and J. Kennedy, Eds., *EuroVis—Short Papers*, The Eurographics Association, 2014. DOI: 10.2312/eurovisshort.20141149 16

N. Kerracher, J. Kennedy, and K. Chalmers. A task taxonomy for temporal graph visualisation. *IEEE Transactions on Visualization and Computer Graphics*, PP(99):1, 2015. DOI: 10.1109/tvcg.2015.2424889 37, 41, 42

A. Kerren and F. Schreiber. Network visualization for integrative bioinformatics. In M. Chen and R. Hofestädt, Eds., *Approaches in Integrative Bioinformatics: Towards the Virtual Cell*, pages 173–202, Springer, 2014. DOI: 10.1007/978-3-642-41281-3_7 29, 88

A. Kerren, H. Purchase, and M. Ward. *Multivariate Network Visualization*, vol. 8380. Springer, 2014a. DOI: 10.1007/978-3-319-06793-3 78, 88

A. Kerren, H. C. Purchase, and M. O. Ward. Introduction to multivariate network visualization. In *Multivariate Network Visualization*, vol. 8380 of *Lecture Notes in Computer Science*, pages 1–9, Springer, 2014b. DOI: 10.1007/978-3-319-06793-3_1 14, 84

M. Kivelä, A. Arenas, M. Barthelemy, J. P. Gleeson, Y. Moreno, and M. A. Porter. Multilayer networks. *Journal of Complex Networks*, 2(3):203–271, 2014. DOI: 10.1093/comnet/cnu016 xi, 1, 2, 4, 5, 6, 12, 15, 16, 21, 25, 26, 31, 33, 68, 72, 82, 83, 97

M. Kivelä, F. McGee, G. Melançon, N. H. Riche, and T. von Landesberger. Visual analytics of multilayer networks across disciplines (Dagstuhl seminar 19061). *Dagstuhl Reports*, 9(2):1–26, 2019. DOI: 10.4230/DagRep.9.2.1 xi

O. Kohlbacher, F. Schreiber, and M. O. Ward. Multivariate networks in the life sciences. In *Multivariate Network Visualization*, vol. 8380 of *Lecture Notes in Computer Science*, pages 61–73, Springer, 2014. DOI: 10.1007/978-3-319-06793-3_4 31, 42, 54

R. N. Kostoff. Science and technology innovation. *Technovation*, 19(10):593–604, 1999. DOI: 10.1016/S0166-4972(99)00084-X 97

J. Kotlarek, O. Kwon, K. Ma, P. Eades, A. Kerren, K. Klein, and F. Schreiber. A study of mental maps in immersive network visualization. In *IEEE Pacific Visualization Symposium (PacificVis)*, pages 1–10, 2020. DOI: 10.1109/pacificvis48177.2020.4722 57, 62

M. Krzywinski, J. Schein, I. Birol, J. Connors, R. Gascoyne, D. Horsman, S. J. Jones, and M. A. Marra. Circos: An information aesthetic for comparative genomics. *Genome Research*, 19(9):1639–1645, 2009. DOI: 10.1101/gr.092759.109 38, 51

M. Krzywinski, I. Birol, S. J. Jones, and M. A. Marra. Hive plots—rational approach to visualizing networks. *Briefings in Bioinformatics*, 13(5):627–644, 2011. DOI: 10.1093/bib/bbr069 38, 50

T.-C. Kuo, T.-F. Tian, and Y. J. Tseng. 3omics: A web-based systems biology tool for analysis, integration and visualization of human transcriptomic, proteomic, and metabolomic data. *BMC Systems Biology*, 7:64, 2013. DOI: 10.1186/1752-0509-7-64 17

M. Kurant and P. Thiran. On survivable routing of mesh topologies in IP-over-WDM networks. In *Proc. 24th Annual Joint Conference of the IEEE Computer and Communications Societies*, 2:1106–1116, 2005. DOI: 10.1109/infcom.2005.1498338 20

M. Kurant and P. Thiran. Layered complex networks. *Physical Review Letters*, 96(13):138701, 2006. DOI: 10.1103/physrevlett.96.138701 20

O. Kwon, C. Muelder, K. Lee, and K. Ma. A study of layout, rendering, and interaction methods for immersive graph visualization. *IEEE Transactions on Visualization and Computer Graphics*, 22(7):1802–1815, July 2016. DOI: 10.1109/tvcg.2016.2520921 57, 62

R. Lambiotte and M. Ausloos. Collaborative tagging as a tripartite network. In *International Conference on Computational Science*, pages 1114–1117, Springer, 2006. DOI: 10.1007/11758532_152 14

M. Latapy, C. Magnien, and N. Del Vecchio. Basic notions for the analysis of large two-mode networks. *Social Networks*, 30(1):31–48, 2008. DOI: 10.1016/j.socnet.2007.04.006 71, 74

A. Laumond, G. Melançon, B. Pinaud, and M. Ghoniem. M-QuBE 3: Querying big multilayer graph by evolutive extraction and exploration. *Journal of Imaging Science and Technology*, 2019. DOI: 10.2352/issn.2470-1173.2019.1.vda-686 34, 66

B. Lavaud-Legendre, C. Plessard, G. Melançon, A. Laumond, and B. Pinaud. Analyse de réseaux criminels de traite des êtres humains: Méthodologie, modélisation et visualisation. *Journal of Interdisciplinary Methodologies and Issues in Science*, 2, 2017. DOI: 10.18713/jimis-300617-2-5 xi, 6

E. Lazega and P. E. Pattison. Multiplexity, generalized exchange and cooperation in organizations: A case study. *Social Networks*, 21:67–90, 1999. DOI: 10.1016/s0378-8733(99)00002-7 2, 6, 18

E. Lazega and T. A. Snijders. *Multilevel network analysis for the social sciences: Theory, methods and applications*, vol. 12, Springer, 2015. DOI: 10.1007/978-3-319-24520-1 7

N. Le Novere, M. Hucka, H. Mi, S. Moodie, F. Schreiber, A. Sorokin, E. Demir, K. Wegner, M. I. Aladjem, S. M. Wimalaratne, et al. The systems biology graphical notation. *Nature Biotechnology*, 27(8):735–741, 2009. DOI: 10.1038/nbt.1558 64

B. Lee, C. Plaisant, C. S. Parr, J.-D. Fekete, and N. Henry. Task taxonomy for graph visualization. In *Proc. AVI Workshop on BEyond Time and Errors: Novel Evaluation Methods for Information Visualization*, pages 1–5, ACM, Venice, Italy, 2006. DOI: 10.1145/1168149.1168168 37, 41, 42

B. Lee, G. Smith, G. G. Robertson, M. Czerwinski, and D. S. Tan. FacetLens: Exposing trends and relationships to support sensemaking within faceted datasets. In *Proc. of the SIGCHI Conference on Human Factors in Computing Systems*, pages 1293–1302, ACM, Boston, MA, 2009. DOI: 10.1145/1518701.1518896 15

F. W. Levi. *Finite Geometrical Systems: Six Public Lectures Delivered in February, 1940, at the University of Calcutta*. The University of Calcutta, 1942. 14

A. Lhuillier, C. Hurter, and A. Telea. State-of-the-art in edge and trail bundling techniques. *Computer Graphics Forum*, 36(3):619–645, 2017. DOI: 10.1111/cgf.13213 63

D. Liben-Nowell and J. Kleinberg. The link-prediction problem for social networks. *Journal of the American Society for Information Science and Technology*, 58(7):1019–1031, 2007. DOI: 10.1002/asi.20591 72

N. Lin. A network theory of social capital. In D. Castiglione, J. W. Van Deth, and G. Wolleb, Eds., *The Handbook of Social Capital*, page 69, Oxford University Press, 2008. 2

X. Liu and H.-W. Shen. The effects of representation and juxtaposition on graphical perception of matrix visualization. In *Proc. of the 33rd Annual ACM Conference on Human Factors in Computing Systems*, pages 269–278, ACM Press, Seoul, Republic of Korea, 2015. DOI: 10.1145/2702123.2702217 47

Y. Liu, C. Wang, P. Ye, and K. Zhang. Hybridvis: An adaptive hybrid-scale visualization of multivariate graphs. *Journal of Visual Languages and Computing*, 41:100–110, 2017. DOI: 10.1016/j.jvlc.2017.03.008 38, 62

T. Luciani, A. Burks, C. Sugiyama, J. Komperda, and G. E. Marai. Details-first, show context, overview last: Supporting exploration of viscous fingers in large-scale ensemble simulations. *IEEE Transactions on Visualization and Computer Graphics*, 25(1):1225–1235, January 2019. DOI: 10.1109/tvcg.2018.2864849 66, 67, 69

D. T. Luu and T. Lux. Multilayer overlaps and correlations in the bank-firm credit network of Spain. *Quantitative Finance*, 19(12):1953–1974, 2019. DOI: 10.1080/14697688.2019.1620318 18

C. Ma, R. V. Kenyon, A. G. Forbes, T. Berger-Wolf, B. J. Slater, and D. A. Llano. Visualizing Dynamic Brain Networks Using an Animated Dual-Representation. In *Eurographics Conference on Visualization (EuroVis)—Short Papers*, EuroVis, The Eurographics Association, 2015. DOI: 10.2312/eurovisshort.20151128 82

J. Mackinlay. Automating the design of graphical presentations of relational information. *ACM Transactions On Graphics (Tog)*, 5(2):110–141, 1986. DOI: 10.1145/22949.22950 45, 46, 63

D. Maier. Minimum covers in relational database model. *Journal of the ACM (JACM)*, 27(4):664–674, 1980. DOI: 10.1145/322217.322223 14

A. Manivannan, W. Q. Yow, R. Bouffanais, and A. Barrat. Are the different layers of a social network conveying the same information? *EPJ Data Science*, 7(1):34, 2018. DOI: 10.1140/epjds/s13688-018-0161-9 2, 7

M. R. Marner, R. T. Smith, B. H. Thomas, K. Klein, P. Eades, and S.-H. Hong. Gion: Interactively untangling large graphs on wall-sized displays. In C. Duncan and A. Symvonis, Eds., *Graph Drawing*, pages 113–124, Springer Berlin Heidelberg, 2014. DOI: 10.1007/978-3-662-45803-7_10 91

K. Marriott, J. Chen, M. Hlawatsch, T. Itoh, M. A. Nacenta, G. Reina, and W. Stuerzlinger. Immersive analytics: Time to reconsider the value of 3D for information visualisation. In *Immersive Analytics*, pages 25–55, Springer, 2018a. DOI: 10.1007/978-3-030-01388-2_2 62

K. Marriott, F. Schreiber, T. Dwyer, K. Klein, N. H. Riche, T. Itoh, W. Stuerzlinger, and B. H. Thomas. *Immersive Analytics*, Springer, Cham, 2018b. DOI: 10.1007/978-3-030-01388-2 57

R. M. Martins, J. F. Kruiger, R. Minghim, A. C. Telea, and A. Kerren. MVN-Reduce: Dimensionality reduction for the visual analysis of multivariate networks. In B. Kozlikova, T. Schreck, and T. Wischgoll, Eds., *EuroVis—Short Papers*, The Eurographics Association, 2017. DOI: 10.2312/eurovisshort.20171126 78

F. McGee and J. Dingliana. An empirical study on the impact of edge bundling on user comprehension of graphs. In *Proc. of the International Working Conference on Advanced Visual Interfaces, AVT'12*, pages 620–627, Association for Computing Machinery, New York, 2012. DOI: 10.1145/2254556.2254670 58

F. McGee, M. During, and M. Ghoniem. Towards visual analytics of multilayer graphs for digital cultural heritage. In *1st Workshop on Visualization for the Digital Humanities*, Baltimore, 2016. https://orbilu.uni.lu/bitstream/10993/31349/1/Towards%20Visual%20Analytics%20of%20Multilayer%20Graphs%20for%20Digital%20Cultural%20Heritage.pdf 6, 19

F. McGee, M. Ghoniem, G. Melançon, B. Otjacques, and B. Pinaud. The state-of-the-art in multilayer network visualization. *Computer Graphics Forum*, 38(6):125–149, 2019a. DOI: 10.1111/cgf.13610 xi, 38

F. McGee, L. Morin, M. Stefas, S. Zorzan, and M. Ghoniem. Layer definition and discovery in multilayer network datasets. In *1st Workshop on the Visualization of Multilayer Networks*, 2019b. https://www.multilayernetvis.org/fileadmin/files/McGee2019LayerDefinintionAndDiscovery.pdf 74

J. M. McPherson. HyperNetwork sampling: Duality and differentiation among voluntary organizations. *Social Networks*, 3(4):225–249, 1982. DOI: 10.1016/0378-8733(82)90001-6 14

M. McPherson, L. Smith-Lovin, and J. M. Cook. Birds of a feather: Homophily in social networks. *Annual Review of Sociology*, 27(1):415–444, 2001. DOI: 10.1146/annurev.soc.27.1.415 17

G. Melançon, H. Ren, and B. Renoust. Handling complex multilayer networks—an approach based on visual network analytics. In *Complex Systems, Smart Territories and Mobility*, pages 51–69, Springer, 2020. DOI: 10.1007/978-3-030-59302-5_3 38, 40, 72

G. Melançon. Just how dense are dense graphs in the real world?: A methodological note. In *Proc. of the AVI Workshop on BEyond Time and Errors: Novel Evaluation Methods for Information Visualization, BELIV'06*, pages 1–7, ACM, Venice, Italy, 2006. DOI: 10.1145/1168149.1168167 21

M. Meyer, M. Sedlmair, and T. Munzner. The four-level nested model revisited: Blocks and guidelines. In *Proc. of the BELIV Workshop: Beyond Time and Errors—Novel Evaluation Methods for Visualization, BELIV'12*, Association for Computing Machinery, 2012. DOI: 10.1145/2442576.2442587 7

J. Moody, D. McFarland, and S. Bender-deMoll. Dynamic network visualization. *American Journal of Sociology*, 110(4):1206–1241, 2005. DOI: 10.1086/421509 16, 54

J. L. Moreno. *Who Shall Survive?: A New Approach to the Problem of Human Interrelations.* Nervous and Mental Disease Publishing Co., 1934. DOI: 10.1037/10648-000 1, 6, 11, 12

T. Moscovich, F. Chevalier, N. Henry, E. Pietriga, and J.-D. Fekete. Topology-aware navigation in large networks. In *Proc. of the SIGCHI Conference on Human Factors in Computing Systems, CHI'09*, pages 2319–2328, ACM, 2009. DOI: 10.1145/1518701.1519056 88

Y. Mourchid, B. Renoust, O. Roupin, L. Văn, H. Cherifi, and M. El Hassouni. MovieNet: A movie multilayer network model using visual and textual semantic cues. *Applied Network Science*, 4(1):1–37, 2019. DOI: 10.1007/s41109-019-0226-0 19, 25

M. Müller, K. Ballweg, T. von Landesberger, S. Yimam, U. Fahrer, C. Biemann, M. Rosenbach, M. Regneri, and H. Ulrich. Guidance for multi-type entity graphs from text collections. In *Proc. of the EuroVis Workshop on Visual Analytics, EuroVA'17*, pages 1–6, Eurographics Association, Goslar Germany, Germany, 2017. DOI: 10.2312/eurova.20171111 19, 58, 134

T. Munzner. A nested process model for visualization design and validation. *IEEE Transactions on Visualization and Computer Graphics*, 15(6):921–928, 2009. DOI: 10.1109/TVCG.2009.111 7

P. Murray, F. McGee, and A. G. Forbes. A taxonomy of visualization tasks for the analysis of biological pathway data. *BMC Bioinformatics*, 18(2):21, February 2017. DOI: 10.1186/s12859-016-1443-5 17, 42

D. Nekrasovski, A. Bodnar, J. McGrenere, F. Guimbretière, and T. Munzner. An evaluation of pan and zoom and rubber sheet navigation with and without an overview. In *Proc. of the SIGCHI Conference on Human Factors in Computing Systems, CHI'06*, pages 11–20, ACM, 2006. DOI: 10.1145/1124772.1124775 91

R. J. Nemeth and D. A. Smith. International trade and world-system structure: A multiple network analysis. *Review (Fernand Braudel Center)*, 8(4):517–560, 1985. http://www.jstor.org/stable/40241006 18

F. F. Nerini, B. Sovacool, N. Hughes, L. Cozzi, E. Cosgrave, M. Howells, M. Tavoni, J. Tomei, H. Zerriffi, and B. Milligan. Connecting climate action with other sustainable development goals. *Nature Sustainability*, 2(8):674–680, 2019. DOI: 10.1038/s41893-019-0334-y 19

S. Nguyen and S. Pallottino. Hyperpaths and shortest hyperpaths. In *Combinatorial Optimization*, pages 258–271, Springer, 1989. DOI: 10.1007/bfb0083470 14

V. Nicosia and V. Latora. Measuring and modeling correlations in multiplex networks. *Physical Review E*, 92(3):032805, 2015. DOI: 10.1103/physreve.92.032805 24

C. Nobre, M. Meyer, M. Streit, and A. Lex. The state-of-the-art in visualizing multivariate networks. *Computer Graphics Forum*, 38(3):807–832, 2019. DOI: 10.1111/cgf.13728 14, 37, 41

C. Nobre, D. Wootton, L. Harrison, and A. Lex. Evaluating multivariate network visualization techniques using a validated design and crowdsourcing approach. In *Proc. of the CHI Conference on Human Factors in Computing Systems, CHI'20*, pages 1–12, Association for Computing Machinery, New York, 2020. DOI: 10.1145/3313831.3376381 61, 91

D. Norman. *The Design of Everyday Things: Revised and Expanded Edition*. Constellation, 2013. 69

M. Okoe, R. Jianu, and S. G. Kobourov. Node-link or adjacency matrices: Old question, new insights. *IEEE Transactions on Visualization and Computer Graphics*, pages 1–1, 2018. DOI: 10.1109/tvcg.2018.2865940 47, 61, 91

E. Omodei, M. D. De Domenico, and A. Arenas. Characterizing interactions in online social networks during exceptional events. *Frontiers in Physics*, 3:59, 2015. DOI: 10.3389/fphy.2015.00059 24

B. Otjacques, M. Noirhomme, and F. Feltz. Taxonomy of the visualization techniques of project related interactions. *Journal of Information Technology in Construction (ITcon)*, 11(41):587–605, 2006. https://itcon.org/papers/2006_41.content.08126.pdf 21

R. A. Pache, A. Céol, and P. Aloy. Netaligner—a network alignment server to compare complexes, pathways and whole interactomes. *Nucleic Acids Research*, 40(W1):W157–W161, 2012. DOI: 10.1093/nar/gks446 41

A. R. Pamfil, S. D. Howison, and M. A. Porter. Inference of edge correlations in multilayer networks. *Physical Review E*, 102(6), December 2020. DOI: 10.1103/physreve.102.062307 28, 72

J. R. Parikh, Y. Xia, and J. A. Marto. Multi-edge gene set networks reveal novel insights into global relationships between biological themes. *PLOS ONE*, 7(9):1–15, September 2012. DOI: 10.1371/journal.pone.0045211 6

P. Parmentier, T. Viard, B. Renoust, and J.-F. Baffier. Introducing multilayer stream graphs and layer centralities. In *International Conference on Complex Networks and their Applications*, pages 684–696, Springer, 2019. DOI: 10.1007/978-3-030-36683-4_55 20

T. Parsons. An analytical approach to the theory of social stratification. *American Journal of Sociology*, 45(6):841–862, 1940. DOI: 10.1086/218489 7

M. C. Pasqual and O. L. de Weck. Multilayer network model for analysis and management of change propagation. *Research in Engineering Design*, 23(4):305–328, 2012. DOI: 10.1007/s00163-011-0125-6 20

R. Pastor-Satorras, C. Castellano, P. Van Mieghem, and A. Vespignani. Epidemic processes in complex networks. *Reviews of Modern Physics*, 87(3):925, 2015. DOI: 10.1103/RevModPhys.87.925 6

G. A. Pavlopoulos, S. I. O'Donoghue, V. P. Satagopam, T. G. Soldatos, E. Pafilis, and R. Schneider. Arena3D: Visualization of biological networks in 3D. *BMC Systems Biology*, 2:104, 2008. DOI: 10.1186/1752-0509-2-104 5, 57

Z. Pei, X. Zhang, F. Zhang, and B. Fang. Attributed multi-layer network embedding. In *IEEE International Conference on Big Data (Big Data)*, pages 3701–3710, 2018. DOI: 10.1109/bigdata.2018.8621900 85

R. Pienta, A. Tamersoy, A. Endert, S. Navathe, H. Tong, and D. H. Chau. Visage: Interactive visual graph querying. In *Proc. of the International Working Conference on Advanced Visual Interfaces*, pages 272–279, 2016. DOI: 10.1145/2909132.2909246 73

W. A. Pike, J. Stasko, R. Chang, and T. A. O'Connell. The science of interaction. *Information Visualization*, 8(4):263–274, 2009. DOI: 10.1057/ivs.2009.22 66

M. Piškorec, B. Sluban, and T. Šmuc. MultiNets: Web-based multilayer network visualization. In *Joint European Conference on Machine Learning and Knowledge Discovery in Databases*, pages 298–302, Springer, 2015. DOI: 10.1007/978-3-319-23461-8_34 68

J. M. Podolny and J. N. Baron. Resources and relationships: Social networks and mobility in the workplace. *American Sociological Review*, pages 673–693, 1997. DOI: 10.2307/2657354 17, 71

M. Pohl and A. Kerren. Human factors and multilayer networks. In *Proc. of the 1st Workshop on Visualization of Multilayer Networks (MNLVIS'19) at IEEE VIS'19*, Vancouver, BC, Canada, October 21, 2019. https://www.multilayernetvis.org/fileadmin/files/Pohl2019HumanFactorsForMultilayerNetworks.pdf 87

P. Pradhan, L. Costa, D. Rybski, W. Lucht, and J. P. Kropp. A systematic study of sustainable development goal (SDG) interactions. *Earth's Future*, 5(11):1169–1179, 2017. DOI: 10.1002/2017ef000632 19

A. J. Pretorius and J. J. Van Wijk. Visual inspection of multivariate graphs. *Computer Graphics Forum*, 27(3):967–974, 2008. DOI: 10.1111/j.1467-8659.2008.01231.x 14, 29, 41, 80, 81, 134

J. Pretorius, H. C. Purchase, and J. T. Stasko. Tasks for multivariate network analysis. In A. Kerren, H. C. Purchase, and M. O. Ward, Eds., *Multivariate Network Visualization: Dagstuhl Seminar #13201*, Dagstuhl Castle, Germany, May 12–17, 2013, Revised Discussions, pages 77–95, Springer International Publishing, 2014. DOI: 10.1007/978-3-319-06793-3_5 37, 39, 41, 42

S. D. S. Reis, Y. Hu, A. Babino, J. S. Andrade, S. Canals, M. Sigman, and H. A. Makse. Avoiding catastrophic failure in correlated networks of networks. *Nature Physics*, 10(10):762–767, September 2014. DOI: 10.1038/nphys3081 20

H. Ren. Visualizing media with interactive multiplex networks. Ph.D. thesis, Bordeaux, 2019. 19

H. Ren, B. Renoust, G. Melançon, M.-L. Viaud, and S. Satoh. Exploring temporal communities in mass media archives. In *Proc. of the 26th ACM International Conference on Multimedia*, pages 1247–1249, 2018a. DOI: 10.1145/3240508.3241392 19

H. Ren, B. Renoust, M. Viaud, G. Melançon, and S. Satoh. Generating "visual clouds" from multiplex networks for TV news archive query visualization. In *International Conference on Content-Based Multimedia Indexing (CBMI)*, pages 1–6, La Rochelle, France, September 2018b. DOI: 10.1109/cbmi.2018.8516482 19, 31

B. Renoust, G. Melançon, and M.-L. Viaud. Measuring group cohesion in document collections. In *IEEE/WIC/ACM International Joint Conferences on Web Intelligence (WI) and Intelligent Agent Technologies (IAT)*, 1:373–380, 2013. DOI: 10.1109/wi-iat.2013.53 xi

B. Renoust, G. Melançon, and M.-L. Viaud. Entanglement in multiplex networks: Understanding group cohesion in homophily networks. In R. Missaoui and I. Sarr, Eds., *Social Network Analysis—Community Detection and Evolution*, Lecture Notes in Social Networks, pages 89–117, Springer, 2014. DOI: 10.1007/978-3-319-12188-8_5 xi, 14, 18, 19, 28

B. Renoust, G. Melançon, and T. Munzner. Detangler: Visual analytics for multiplex networks. *Computer Graphics Forum*, 34(3):321–330, 2015. DOI: 10.1111/cgf.12644 xi, 5, 6, 24, 26, 28, 29, 31, 33, 34, 38, 39, 40, 56, 58, 69, 70, 72, 75, 80, 83, 88, 98, 99, 133, 134

B. Renoust, D.-D. Le, and S. Satoh. Visual analytics of political networks from face-tracking of news video. *IEEE Transactions on Multimedia*, 18(11):2184–2195, 2016. DOI: 10.1109/tmm.2016.2614224 19

B. Renoust, M. Oliveira Franca, J. Chan, V. Le, A. Uesaka, Y. Nakashima, H. Nagahara, J. Wang, and Y. Fujioka. Buda.art: A multimodal content based analysis and retrieval system for buddha statues. In *Proc. of the 27th ACM International Conference on Multimedia*, pages 1062–1064, 2019a. DOI: 10.1145/3343031.3350591 19

B. Renoust, H. Ren, and G. Melançon. Animated drag and drop interaction for dynamic multidimensional graphs. *ArXiv Preprint ArXiv:1902.01564*, 2019b. https://arxiv.org/abs/1902.01564 40, 70, 71, 134

B. Renoust, H. Ren, G. Melançon, M.-L. Viaud, and S. Satoh. A multimedia document browser based on multilayer networks. *Multimedia Tools and Applications*, pages 1573–7721, 2020. DOI: 10.1007/s11042-020-09872-9 31

L. Rossi and M. Magnani. Towards effective visual analytics on multiplex and multilayer networks. *Chaos, Solitons and Fractals*, 72:68–76, Multiplex Networks: Structure, Dynamics and Applications, 2015. DOI: 10.1016/j.chaos.2014.12.022 8, 75, 78, 90

M. Rosvall and C. T. Bergstrom. Mapping change in large networks. *PloS One*, 5(1), 2010. DOI: 10.1371/journal.pone.0008694 16, 34, 35, 133

S. Rufiange and G. Melançon. Animatrix: A matrix-based visualization of software evolution. In *2nd IEEE Working Conference on Software Visualization*, pages 137–146, 2014. DOI: 10.1109/vissoft.2014.30 82

S. Rufiange, M. J. McGuffin, and C. P. Fuhrman. Treematrix: A hybrid visualization of compound graphs. *Computer Graphics Forum*, 31(1):89–101, 2012. DOI: 10.1111/j.1467-8659.2011.02087.x 62

J. Sansen, R. Bourqui, B. Pinaud, and H. Purchase. Edge visual encodings in matrix-based diagrams. In *19th International Conference on Information Visualisation (IV)*, pages 62–67, 2015. https://hal.archives-ouvertes.fr/hal-01189166 DOI: 10.1109/iv.2015.22 62, 91

R. Santamaría, R. Therón, and L. Quintales. A visual analytics approach for understanding biclustering results from microarray data. *BMC Bioinformatics*, 9:247, 2008. DOI: 10.1186/1471-2105-9-247 38, 47, 56, 63

A. Saumell-Mendiola, M. Á. Serrano, and M. Boguná. Epidemic spreading on interconnected networks. *Physical Review E*, 86(2), 2012. DOI: 10.1103/physreve.86.026106 6

F. Schreiber, A. Kerren, K. Börner, H. Hagen, and D. Zeckzer. Heterogeneous networks on multiple levels. In A. Kerren, H. C. Purchase, and M. O. Ward, Eds., *Multivariate Network Visualization: Dagstuhl Seminar #13201*, Dagstuhl Castle, Germany, May 12–17, 2013, Revised Discussions, pages 175–206, Springer International Publishing, 2014. DOI: 10.1007/978-3-319-06793-3_9 6, 14, 88, 134

R. Shadoan and C. Weaver. Visual analysis of higher-order conjunctive relationships in multidimensional data using a hypergraph query system. *IEEE Transactions on Visualization and Computer Graphics*, 19(12):2070–2079, 2013. DOI: 10.1109/tvcg.2013.220 31

C. E. Shannon. A mathematical theory of communication. *The Bell System Technical Journal*, 27:379–423, 623–656, 1948. DOI: 10.1002/j.1538-7305.1948.tb01338.x 80

L. M. Shekhtman, M. M. Danziger, and S. Havlin. Recent advances on failure and recovery in networks of networks. *Chaos, Solitons and Fractals*, 90:28–36, 2016. DOI: 10.1016/j.chaos.2016.02.002 20

Z. Shen, K.-L. Ma, and T. Eliassi-Rad. Visual analysis of large heterogeneous social networks by semantic and structural abstraction. *IEEE Transactions on Visualization and Computer Graphics*, 12(6):1427–1439, November 2006. DOI: 10.1109/tvcg.2006.107 29, 38, 41, 68, 80

L. Shi, Q. Liao, H. Tong, Y. Hu, Y. Zhao, and C. Lin. Hierarchical focus+context heterogeneous network visualization. In *IEEE Pacific Visualization Symposium*, pages 89–96, Yokohama, Japan, March 2014. DOI: 10.1109/PacificVis.2014.44 13, 33, 88

L. Shi, Q. Liao, H. Tong, Y. Hu, C. Wang, C. Lin, and W. Qian. OnionGraph: Hierarchical topology+attribute multivariate network visualization. *Visual Informatics*, 2020. DOI: 10.1016/j.visinf.2020.01.002 68, 69, 134

B. Shneiderman. The eyes have it: A task by data type taxonomy for information visualizations. In *Proc. of the IEEE Symposium on Visual Languages*, pages 336–343, IEEE Computer Society Press, 1996. DOI: 10.1109/VL.1996.545307 65, 67, 69

B. Shneiderman and A. Aris. Network visualization by semantic substrates. *IEEE Transactions on Visualization and Computer Graphics*, 12(5):733–740, 2006. DOI: 10.1109/tvcg.2006.166 85

B. Škrlj and B. Renoust. Layer entanglement in multiplex, temporal multiplex, and coupled multilayer networks. *Applied Netwrok Science*, 5(89), 2020. DOI: 10.1007/s41109-020-00331-w 17, 18, 21, 25, 72

B. Skrlj, J. Kralj, and N. Lavrac. Py3plex toolkit for visualization and analysis of multilayer networks. *Applied Network Science*, 4(94), 2019. DOI: 10.1007/s41109-019-0203-7 24, 28, 54, 72

B. Sluban, M. Grčar, and I. Ozetič. Temporal multi-layer network construction from major news events. In H. Cherifi, B. Gonçalves, R. Menezes, and R. Sinatra, Eds., *Proc. of the 7th Workshop on Complex Networks CompleNet*, pages 29–41, Springer International Publishing, Dijon, France, 2016. DOI: 10.1007/978-3-319-30569-1_3 6, 19

G. Ślusarczyk, A. Łachwa, W. Palacz, B. Strug, A. Paszyńska, and E. Grabska. An extended hierarchical graph-based building model for design and engineering problems. *Automation in Construction*, 74:95–102, 2017. DOI: 10.1016/j.autcon.2016.11.008 5, 20

G. Smith, M. Czerwinski, B. R. Meyers, G. Robertson, and D. S. Tan. FacetMap: A scalable search and browse visualization. *IEEE Transactions on Visualization and Computer Graphics*, 12(5):797–804, 2006. DOI: 10.1109/tvcg.2006.142 15

D. Snyder and E. L. Kick. Structural position in the world system and economic growth, 1955–1970: A multiple-network analysis of transnational interactions. *American Journal of Sociology*, 84(5):1096–1126, 1979. DOI: 10.1086/226902 12, 18

A. Srinivasan, H. Park, A. Endert, and R. C. Basole. Graphiti: Interactive specification of attribute-based edges for network modeling and visualization. *IEEE Transactions on Visualization and Computer Graphics*, 24(1):226–235, January 2018. DOI: 10.1109/tvcg.2017.2744843 33, 73

J. Stasko, C. Görg, and Z. Liu. Jigsaw: Supporting investigative analysis through interactive visualization. *Information Visualization*, 7(2):118–132, 2008. DOI: 10.1057/palgrave.ivs.9500180 38, 40, 50, 56

J. Sweller. Cognitive load during problem solving: Effects on learning. *Cognitive Science*, 12(2):257–285, 1988. DOI: 10.1207/s15516709cog1202_4 92, 94

J. Sweller, P. Ayres, and S. Kalyuga. *Cognitive Load Theory*. Explorations in the Learning Sciences, Instructional Systems and Performance Technologies, Springer New York, 2011. DOI: 10.1007/978-1-4419-8126-4 92, 94

R. Tamassia, Ed. *Handbook on Graph Drawing and Visualization*. Chapman and Hall/CRC, 2013. https://www.crcpress.com/Handbook-of-Graph-Drawing-and-Visualization/Tamassia/9781584884125 DOI: 10.1201/b15385 88

B. H. Thomas, G. F. Welch, P. Dragicevic, N. Elmqvist, P. Irani, Y. Jansen, D. Schmalstieg, A. Tabard, N. A. ElSayed, R. T. Smith, et al. Situated analytics. *Immersive Analytics*, 11190:185–220, 2018. DOI: 10.1007/978-3-030-01388-2_7 62

T. Thorpe and S. Mead. Project-specific web sites: Friend or foe? *Journal of Construction Engineering and Management*, 127(5):406–413, 2001. DOI: 10.1061/(asce)0733-9364(2001)127:5(406) 20

A. S. O. Toledo, L. C. Carpi, and A. P. F. Atman. Diversity analysis exposes unexpected key roles in multiplex crime networks. In *Complex Networks XI*, pages 371–382, Springer, 2020. DOI: 10.1007/978-3-030-40943-2_31 18

Y. Tu and H. W. Shen. GraphCharter: Combining browsing with query to explore large semantic graphs. In *IEEE Pacific Visualization Symposium (PacificVis)*, pages 49–56, Sydney, NSW, Australia, 2013. DOI: 10.1109/pacificvis.2013.6596127 31

S. Van den Elzen and J. J. Van Wijk. Multivariate network exploration and presentation: From detail to overview via selections and aggregations. *IEEE Transactions on Visualization and Computer Graphics*, 20(12):2310–2319, 2014. DOI: 10.1109/TVCG.2014.2346441 66, 74

F. Van Ham. Using multilevel call matrices in large software projects. In *IEEE Symposium on Information Visualization*, pages 227–232, Seattle, WA, 2003. DOI: 10.1109/infvis.2003.1249030 47

F. Van Ham and A. Perer. Search, show context, expand on demand: Supporting large graph exploration with degree-of-interest. *IEEE Transactions on Visualization and Computer Graphics*, 15(6):953–960, November 2009. DOI: 10.1109/tvcg.2009.108 66

I. Van Vugt. Using multi-layered networks to disclose books in the republic of letters. *Journal of Historical Network Research*, 1(1):25–51, October 2017. DOI: 10.25517/jhnr.v1i1.7 5, 19

C. Vehlow, F. Beck, P. Auwärter, and D. Weiskopf. Visualizing the evolution of communities in dynamic graphs. *Computer Graphics Forum*, 34(1):277–288, 2015. DOI: 10.1111/cgf.12512 38, 41

C. Vehlow, F. Beck, and D. Weiskopf. Visualizing group structures in graphs: A survey. *Computer Graphics Forum*, 36(6):201–225, 2017. DOI: 10.1111/cgf.12872 61, 88

L. M. Verbrugge. Multiplexity in adult friendships. *Social Forces*, 57(4):1286–1309, 1979. DOI: 10.1093/sf/57.4.1286 1

C. Viau, M. J. McGuffin, Y. Chiricota, and I. Jurisica. The flowvizmenu and parallel scatterplot matrix: Hybrid multidimensional visualizations for network exploration. *IEEE Transactions on Visualization and Computer Graphics*, 16(6):1100–1108, 2010. DOI: 10.1109/tvcg.2010.205 79

R. Vinuesa, H. Azizpour, I. Leite, M. Balaam, V. Dignum, S. Domisch, A. Felländer, S. D. Langhans, M. Tegmark, and F. F. Nerini. The role of artificial intelligence in achieving the sustainable development goals. *Nature Communications*, 11(1):1–10, 2020. DOI: 10.1038/s41467-019-14108-y 19

A. Vogogias, D. Archambault, B. Bach, and J. Kennedy. Visual encodings for networks with multiple edge types. In *Advanced Visual Interfaces (ACM AVI)*, 2020. DOI: 10.1145/3399715.3399827 48, 61, 91

T. von Landesberger, A. Kuijper, T. Schreck, J. Kohlhammer, J. Van Wijk, J.-D. Fekete, and D. Fellner. Visual analysis of large graphs: State-of-the-art and future research challenges. *Computer Graphics Forum*, 30(6):1719–1749, 2011. DOI: 10.1111/j.1467-8659.2011.01898.x 88, 89

Y. Wang and G. Xiao. Epidemics spreading in interconnected complex networks. *Physics Letters A*, 376(42-43):2689–2696, 2012. DOI: 10.1016/j.physleta.2012.07.037 6

C. Ware and R. Bobrow. Supporting visual queries on medium-sized node-link diagrams. *Information Visualization*, 4(1):49–58, 2005. DOI: 10.1057/palgrave.ivs.9500090 90

C. Ware and P. Mitchell. Visualizing graphs in three dimensions. *ACM Transactions on Applied Perception (TAP)*, 5(1):2:1–2:15, January 2008. DOI: 10.1145/1279640.1279642 57

M. Wattenberg. Visual exploration of multivariate graphs. In *Proc. of the SIGCHI Conference on Human Factors in Computing Systems*, pages 811–819, ACM, Montréal, Québec, Canada, 2006. DOI: 10.1145/1124772.1124891 29, 38

K. Wehmuth, E. Fleury, and A. Ziviani. On multiaspect graphs. *Theoretical Computer Science*, 651:50–61, 2016. DOI: 10.1016/j.tcs.2016.08.017 5, 35

K. Wehmuth, É. Fleury, and A. Ziviani. Multiaspect graphs: Algebraic representation and algorithms. *Algorithms*, 10(1):1, 2017. DOI: 10.3390/a10010001 35

S. White and S. Feiner. Sitelens: Situated visualization techniques for urban site visits. In *Proc. of the SIGCHI Conference on Human Factors in Computing Systems*, pages 1117–1120, Association for Computing Machinery, 2009. DOI: 10.1145/1518701.1518871 62

L. Wilkinson and M. Friendly. The history of the cluster heat map. *The American Statistician*, 63(2):179–184, 2009. DOI: 10.1198/tas.2009.0033 47

J. Xia, E. E. Gill, and R. E. Hancock. Networkanalyst for statistical, visual and network-based meta-analysis of gene expression data. *Nature Protocols*, 10(6):823–44, 2015. DOI: 10.1038/nprot.2015.052 38, 52

H. Yang, K. Tang, X. Liu, L. Xiao, R. Xu, and S. Kumara. A user-centred approach to information visualisation in nano-health. *International Journal of Bioinformatics Research and Applications*, 12(2):95–115, 2016. DOI: 10.1504/ijbra.2016.077122 50, 51, 133

J. S. Yi, Y. Ah Kang, and J. Stasko. Toward a deeper understanding of the role of interaction in information visualization. *IEEE Transactions on Visualization and Computer Graphics*, 13(6):1224–1231, 2007. DOI: 10.1109/tvcg.2007.70515 66, 67

J. S. Yi, N. Elmqvist, and S. Lee. TimeMatrix: Analyzing temporal social networks using interactive matrix-based visualizations. *International Journal of Human-Computer Interaction*, 26(11–12):1031–1051, 2010. DOI: 10.1080/10447318.2010.516722 82

V. Yoghourdjian, D. Archambault, S. Diehl, T. Dwyer, K. Klein, H. C. Purchase, and H.-Y. Wu. Exploring the limits of complexity: A survey of empirical studies on graph visualisation. *Visual Informatics*, 2(4):264–282, 2018. DOI: 10.1016/j.visinf.2018.12.006 89, 93

V. Yoghourdjian, T. Dwyer, K. Klein, K. Marriott, and M. Wybrow. Graph thumbnails: Identifying and comparing multiple graphs at a glance. *IEEE Transactions on Visualization and Computer Graphics*, 24(12):3081–3095, December 2018. DOI: 10.1109/tvcg.2018.2790961 31, 32, 90, 133

A. Zeng and S. Battiston. The multiplex network of EU lobby organizations. *PLOS One*, 11(10):e0158062, 2016. DOI: 10.1371/journal.pone.0158062 33, 54

J. Zhao, C. Collins, F. Chevalier, and R. Balakrishnan. Interactive exploration of implicit and explicit relations in faceted datasets. *IEEE Transactions on Visualization and Computer Graphics*, 19(12):2080–2089, 2013. DOI: 10.1109/tvcg.2013.167 15

B. Zimmer, M. Sahlgren, and A. Kerren. Visual analysis of relationships between heterogeneous networks and texts: An application on the IEEE vis publication dataset. *Informatics*, 4(2), 2017. DOI: 10.3390/informatics4020011 68, 80, 88

M. M. Zloof. Query-by-example: A data base language. *IBM Systems Journal*, 16(4):324–343, 1977. DOI: 10.1147/sj.164.0324 73

Authors' Biographies

FINTAN MCGEE

Dr. Fintan McGee is a research associate at the Luxembourg Institute of Science and Technology (LIST). He received his Ph.D. in Computer Science from Trinity College Dublin, in 2013. He was an organizer of the Dagstuhl Seminar on "Visual Analytics of Multilayer Networks across Disciplines" (#19061) as well as the Multilayer Network visualization workshop at VIS 2019. He was work package leader on the BLIZAAR project, an international collaboration, that focuses on multilayer network visualization. His research focuses on visualization of biological data (a key application domain for multilayer network visualization), including multivariate analytics, as well as visualization evaluation.

BENJAMIN RENOUST

Dr. Benjamin Renoust is a guest associate professor at the Osaka University, Institute for Datability Science since 2017, and senior data scientist at Median Technologies since 2019. He is also a visiting lecturer at the National Institute of Informatics (NII), and at the CNRS UMI 3527 Japanese–French Laboratory for Informatics (JFLI), Japan (since 2014). He was a research engineer at the National Audiovisual Institute (Ina) in Paris, France, from 2009–2012, and received his Ph.D. in 2014 from the University of Bordeaux, France (with a Ph.D. thesis dedicated to the visualization and analysis of multiplex networks). He cofounded in 2019 the French chapter of the Complex System Society. His research is focused on network visual analytics and media analytics, with applications in a large variety of domains ranging from humanities and medical imagery, to law and quantum physics. Benjamin especially uses multilayer networks to interact with the heterogeneity of data in each of these domains.

DANIEL ARCHAMBAULT

Dr. Daniel Archambault received his Ph.D. in Computer Science from the University of British Columbia, Canada in 2008. He is an Associate Professor of Computer Science at Swansea University in the United Kingdom. His principle area of research is the scalable interactive visualization of networks in both static and dynamic settings. He also has interests in many areas of information visualization, visual analytics, text analysis and visualization, social media analytics, visual data science, and perceptual factors in visualization.

MOHAMMAD GHONIEM

Dr. Mohammad Ghoniem is a senior research and technology associate at the Luxembourg Institute of Science and Technology. He received his doctorate in Computer Science from the University of Nantes, France, in 2005. His main research interests include information visualization, visual analytics, and usability evaluation of information visualization. Over the past decade, Dr. Ghoniem has been involved, including as a PI, in multiple research projects at the intersection of information visualization and various application domains such as financial fraud detection, retail analytics, broadcast media analytics, metagenomics and network security. He was the Luxembourg-based PI for the BLIZAAR international collaborative research project, dedicated to the visualization of multilayer networks (2016–2019).

ANDREAS KERREN

Prof. Dr. Andreas Kerren received his Ph.D. degree in Computer Science from Saarland University, Saarbrücken (Germany). In 2008, he achieved his habilitation (docent competence) from Växjö University. Dr. Kerren joined Linköping University as a Full Professor in 2020 and holds the chair of Information Visualization. He is also a Full Professor (part time) in Computer Science at the Department of Computer Science and Media Technology, Linnaeus University, where he heads the research group for Information and Software Visualization, called ISOVIS. His main research interests include the areas of Information Visualization, Visual Analytics, and Human-Computer Interaction. He is, among others, an editorial board member of the *Information Visualization* and *Computer Graphics Forum* journals, has served as organizer/program chair at various conferences, such as IEEE VISSOFT 2013/2018, GD 2018, or IVAPP 2013–15/2018–20, and has edited a number of successful books on human-centered visualization.

BRUNO PINAUD

Dr. Bruno Pinaud received his Ph.D. in Computer Science at University of Nantes, France in 2006. Since 2008, he is an associate professor in Computer Science at University of Bordeaux, France. He received his habilitation (HDR in French) in October 2019. His work has focused on visual analytics, graph rewriting systems modeling and visualization, and experimental evaluation. He was the French PI on the BLIZAAR project, an international collaboration, that focuses on multilayer network visualization between researchers from France and Luxembourg. He is a co-author of a recent survey on multilayer network visualization published in Computer Graphics Forum. Bruno Pinaud is also an active developer of the Tulip information visualization framework.

MARGIT POHL

Dr. Margit Pohl received her Master's degree in Computer Science at University of Vienna in 1989 and her Ph.D. in Psychology in 1994 at University of Vienna. She received her habilitation at Vienna University of Technology in 2002. Since 2001 she has been an associate professor at Vienna University of Technology. Her main research interests are HCI and visualization, visual analytics, and cognitive psychology. In this context, she has especially addressed issues of reasoning with complex visualizations. She has published extensively on the appropriate design of complex visualizations. She has co-organized several workshops and tutorials in the area (e.g., EuroVA Symposium, tutorial on comparative visualization at VIS 2018).

BENOÎT OTJACQUES

Dr. Ir. Benoît Otjacques leads the Data Science and Analytics Unit of LIST focused on data analytics, visualization, and interactive technologies. He has been working in visualization for more than 15 years. Beside his research work, he has also been leading various projects involving private companies and public administration. He holds a Ph.D. in Computer Science from the University of Namur (Belgium), an Engineering degree from Ecole Polytechnique de Louvain (Belgium), as well as a University Certificate in Business Ethics and Compliance Management from Louvain School of Management. He has served as organizer/chair of several conferences, like CDVE 2009, EMISA 2014, EGC 2015, and EnviroInfo 2017.

GUY MELANÇON

Prof. Guy Melançon's primary research area is network visual analytics, focusing specifically on multilayer and dynamic networks, with extensive experience in multi-disciplinary research projects. The application domains of his work have included human trafficking networks, digital ethnography, historical archives, spatial geography, and software engineering. He was an organizer of the recent Dagstuhl Seminar on "Visual Analytics of Multilayer Networks across Disciplines" (#19061), keynote speaker at the multilayer network visualization workshop at VIS 2019, and has recently been made a vice-president at the University of Bordeaux.

TATIANA VON LANDESBERGER

Prof. Tatiana von Landesberger is the professor of visualization and Visual Analytics at University of Cologne, Germany previously she has been a professor of Visual Analytics at Universität Rostock, Germany. Her research focuses on visual analytics of networks and spatio-temporal data. She has experience with visual exploration of multilayer networks in the journalism domain as well as network visualization in biology, finance, and medicine. She serves on program committees of visualization research conferences. She was an organizer of the recent Dagstuhl Seminar on "Visual Analytics of Multilayer Networks across Disciplines" (#19061). She has been involved in the organization of several visualization workshops (e.g., VMV, EuroVA, as well as the Multilayer Network visualization workshop at VIS 2019), panel, and tutorials (e.g., at VIS 2014 and 2018).

List of Figures